Saab 35 Draken

Written by Mikhail Putnikov

Walk Around

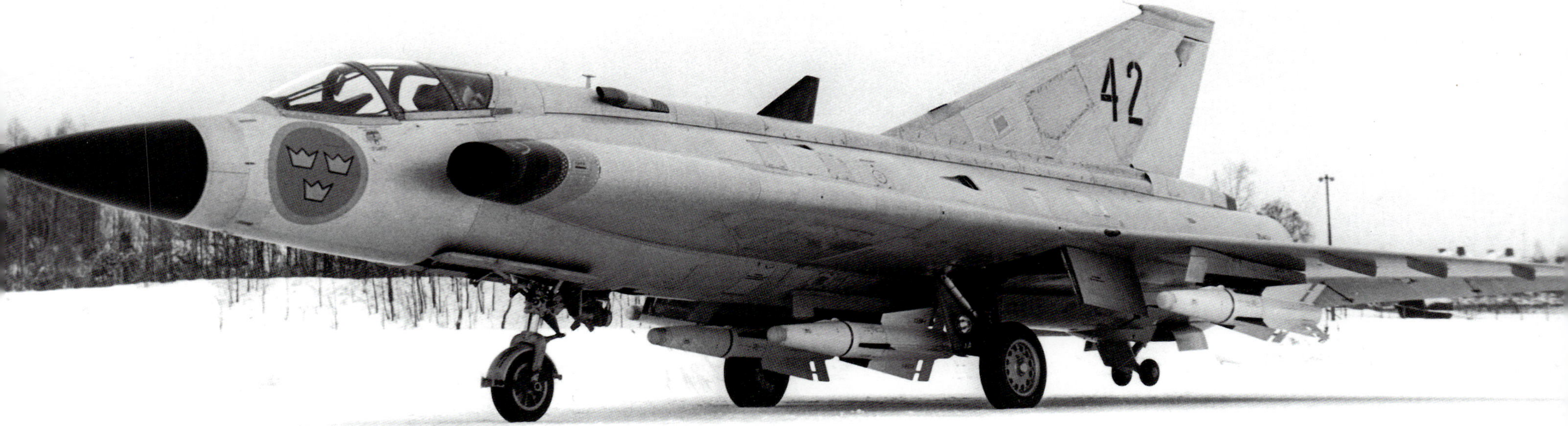

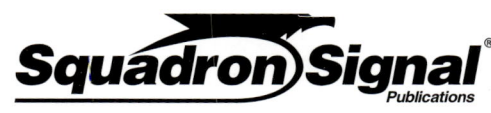

Cover Art by Don Greer
Profiles by Matheu Spraggins
Line Illustrations by Melinda Turnage

Squadron/Signal Publications

(Front Cover) Saab J35J 10-13 fires an Rb 27 (Falcon) radar homing-missile. Missile armament and a fire-control system integrated with a highly advanced semi-automatic air-defense system constituted the effective weapon system that defended Scandinavian skies during the Cold War years.

(Back Cover) A unique front view shows the main construction features of the J 35J: the single gun blast port, Hughes IR scanner, additional communication antenna, and extra hardpoints below the engine intakes.

About the Walk Around®/On Deck Series®

The Walk Around®/On Deck® series is about the details of specific military equipment using color and black-and-white archival photographs and photographs of in-service, preserved, and restored equipment. *Walk Around®* titles are devoted to aircraft and military vehicles, while *On Deck®* titles are devoted to warships. They are picture books of 80 pages, focusing on operational equipment, not one-off or experimental subjects.

Copyright 2010 Squadron/Signal Publications
1115 Crowley Drive, Carrollton, TX 75006-1312 U.S.A.
Printed in the U.S.A.

All rights reserved. No part of this publication may be reproduced, stored in a retrieval system, or transmitted in any form by means electrical, mechanical, or otherwise, without written permission of the publisher.

ISBN 978-0-89747-612-6

Military/Combat Photographs and Snapshots

If you have any photos of aircraft, armor, soldiers, or ships of any nation, particularly wartime snapshots, please share them with us and help make Squadron/Signal's books all the more interesting and complete in the future. Any photograph sent to us will be copied and returned. Electronic images are preferred. The donor will be fully credited for any photos used. Please send them to:

Squadron/Signal Publications
1115 Crowley Drive
Carrollton, TX 75006-1312 U.S.A.
www.SquadronSignalPublications.com

Acknowledgments

Among the many people who helped with this project special gratitude goes to the staff of the Flygvapnet Museum of Linköping, Swedish Aviation Society (SFF); Swedish Military Archives; Finnish Aviation Museum, Vantaa, Helsinki; Aviation Museum of Central Finland, Tikkakoski; and the staff of the Russian magazine M-Hobby. Sincere thanks go to Mr. Jyrki Laukkanen who advised me and shared his photos with me, to Uno Andersson and Sven-Erik Jonsson from SFF, to Stellan Englund (IPMS Sweden), Mike Reeves (IPMS USA), and all my friends for their confidence and support.

For their generosity and willingness to share their photos, I am indebted to: Anders Nylén, Bertil Olofsson, Frank Grealish, Jan Jorgensen, Ingemar Eliasson, Luc Colin, Alan Radecki, Braccini Riccardo, Sven Stauffer, Lars Soldéus, Åke Andersson, Bo Widfeldt, I Thuresson, Rune Rydh, Miroslav Patočka, Rainer Mueller, Joop de Groot, Johann Janschitz, Thomas Welander, Karl Drage, and Markus Willmann.

Introduction

The history of the Saab 35 Draken program can be traced back to the mid-1950s when the Flygvapnet (Swedish Air Force) issued new fighter requirements for the replacement of the J 29 Tunnan and A/J 32 Lansen, two earlier Saab designs that helped to bring Sweden into the jet age. The Saab 35, one of the country's most famous fighters, was born when the Cold War between East and West was raging, and it was very much a product of its time. Led by designer Erik Bratt, Saab's engineering team created an effective weapon that defended Scandinavian skies during the long Cold War years. Many different versions were built to perform a wide range of tasks. The J 35A, J 35B, J 35D, J 35F, and J 35J were fighter versions, with J 35J remaining in Flygvapnet service until 1998. Also developed was a reconnaissance version, the S 35E, which was equipped with seven fixed cameras. In order to facilitate conversion training of new Draken pilots, a two-seater version was also developed under the designation SK 35C.

Designed solely with the Swedish national interest in mind, the Saab 35 has nevertheless achieved considerable success in the export market. The aircraft's excellent performance resulted in long service with the Swedish, Finnish, Austrian, and Danish Air Forces. The Danish Air Force chose a modified version of the Draken in order to fulfill its strike aircraft requirement. The different versions of this variant, known as the F-35, RF-35, and the TF-35, served with the Danish Air Force between 1970 and 1993 and proved very satisfactory. The Draken served the Austrian and Finnish air forces in its original interceptor role. Finland retired the Draken in the year 2000. Austria was the last country to operate the Draken, retaining the J 35D in service until 2005. The official ceremony commemorating the last Draken flight as an operational combat aircraft with the Austrian airforce ended the Draken's long career. In all, a total of 615 Drakens were built, including the prototypes and test aircraft. The type remained in active service for over 40 years. By 1999 the accumulated flight time racked up by Swedish, Danish, Austrian, and Finnish Drakens exceeded 1 million hours.

The pioneering tailless double-delta design proved amazingly successful and ushered in a whole new line of more sophisticated successors. Almost gone but not to be forgotten, Saab's remarkable Draken represented a key stage in maintaining Sweden's seat at the top of the table of fighter producers – a position achieved with J 29 Tunnan and continuing today with Sweden's current JAS 39C/D Gripen.

Perhaps it should be left to the pilots to pronounce the final verdict on the Draken: "Anyone who has ever had the chance to fly the Saab 35 aircraft can only be impressed by the magnitude of this jet, which is called a 'fighter pilot's dream.'"

Now, rather than repeat all dry statistics on the Saab 35 found in numerous other publications, we would like to take you on a walk-around tour and show you every detail of the magnificent Saab 35 Draken.

(Title Page) The first J 35F prototype was converted from the early-production J 35A, s/n 35082 airframe and was used to test armament and electronic systems from 1965 to 1967. It has a flat canopy, extended engine intake lips, and new auxiliary air intake on the dorsal spine. (Bertil Olofsson, Swedish Military Archives)

This photograph, taken in Linköping, Sweden, on 21 January 1952, shows the Saab 210 original engine air intake configuration. There are no doors to cover the landing gear bays. (SFF Photo archive)

The Saab 210A undergoes a test flight in May 1952. Tufting has been applied to the airplane's starboard wing in order to facilitate the recording of airflow patterns over the wing. The landing gear was not fully retractable and was partially exposed during flight. (SFF Photo archive)

Saab 210

Initial discussion on a new fighter to replace the J 29 Tunnan in the Flygvapnet began in the early autumn of 1949. The new aircraft had to intercept Soviet bombers flying at altitudes of some 10,000m at speeds estimated to be in the transonic range (Mach 0.9). Saab engineers not only had to produce a supersonic fighter armed with air-to-air missiles but also one that met the unique requirements of Flygvapnet's operational doctrine, in particular, the ability to fly from dispersed air bases, using strengthened and widened sections of public roads. The aircraft's operational turnaround time also had to be under 10 minutes.

In November 1949 Saab engineers under designer Erik Bratt examined a number of options to meet the specification demands. According to one Saab designer, "the whole aircraft was built around the combination of the powerful engine and radar." Bratt came up with a unique double-delta wing. Its 80°-swept inner wing was thick enough to accommodate the air intakes, weapons, fuel, and landing gears, while the thinner outer portion had a 60° sweep and provided lift for short-field and low-speed operations while keeping drag low enough for supersonic flight. This configuration was chosen for what was designated "Project 1250."

To evaluate the proposed design, particularly the behavior of the new wing at lower speeds, a test aircraft was built: the double-delta prototype listed as the Saab 210 but unofficially dubbed "Lill Draken" (Little Dragon), powered by an Armstrong Siddeley Adder jet engine.

Intensive flight testing began when Saab's chief test pilot, Bengt Olow, took "Lill Draken" up for its first, 25-minute flight on 21 January 1952. The double-delta configuration proved excellent for both high-speed flight and for relatively low take-off and landing speeds. Testing continued as various pilots took the Saab 210 on 887 flights, chalking up 286 flying hours.

The "Lill-Draken" changed its look three times. The Saab 210A had air inlets that extended further towards the nose whereas the 210B had inlets that were drawn further back starting in line with the cockpit. Some minor changes were also made to the wing.

Unarmed Draken prototypes, 35-1 and 35-2, with very short and angled tail cones, serve as flight test airframes in March 1956. (SFF Photo archive)

Draken prototype 35-1 "Röd Urban" (Red U) is seen at the Saab facility in Linköping, on 14 December 1955. (SFF Photo archive)

Prototypes

Bengt Olow was at the controls of the 35-1 Draken prototype when it made its first test flight on 25 October 1955. This aircraft had a British Rolls Royce Avon Mk 21 engine without afterburner, built by Svenska Flygmotor as the RM5A. Later the 35-1 prototype was re-engined with the more powerful Avon Mk 43 without afterburner. With this improved engine, the aircraft flew past Mach 1 in level flight on 26 January 1956. Unfortunately the first prototype suffered damage in a belly landing on 19 April that year and spent several months in the repair shop.

Prototype, 35-2, with an even more powerful Avon Mk 46 engine first took to the air on March 1956, but was also damaged in a belly landing a week before the first prototype's accident.

In September 1956 aircraft 35-3 joined the flight test program. Powered by Avon Mk 46 engine with an afterburner, it was the first cannon-armed Draken but still lacked radar equipment.

Another two prototypes, the 35-4 and 35-5, were built with J 35A features and flew in 1958. Draken 35-4 had a modified canopy and tail cone and RM6B engine – the first license-built Avon Mk 48 engine. Draken 35-5 was the first true production-standard test aircraft for the J 35A series.

The following prototypes - 35-6, 35-7, 35-8, 35-9, 35-10, 35-11, 35-12 and 35-13 - were used for developing new aircraft variants, avionics, and weapons testing. With the production of the J 35A variant, Draken began its years of service in the Swedish, Finnish, Austrian and Danish air forces. The aircraft's successful design allowed for the creation of a multi-role combat aircraft system, which retained in service until 2005.

Saab 35 Production Variants

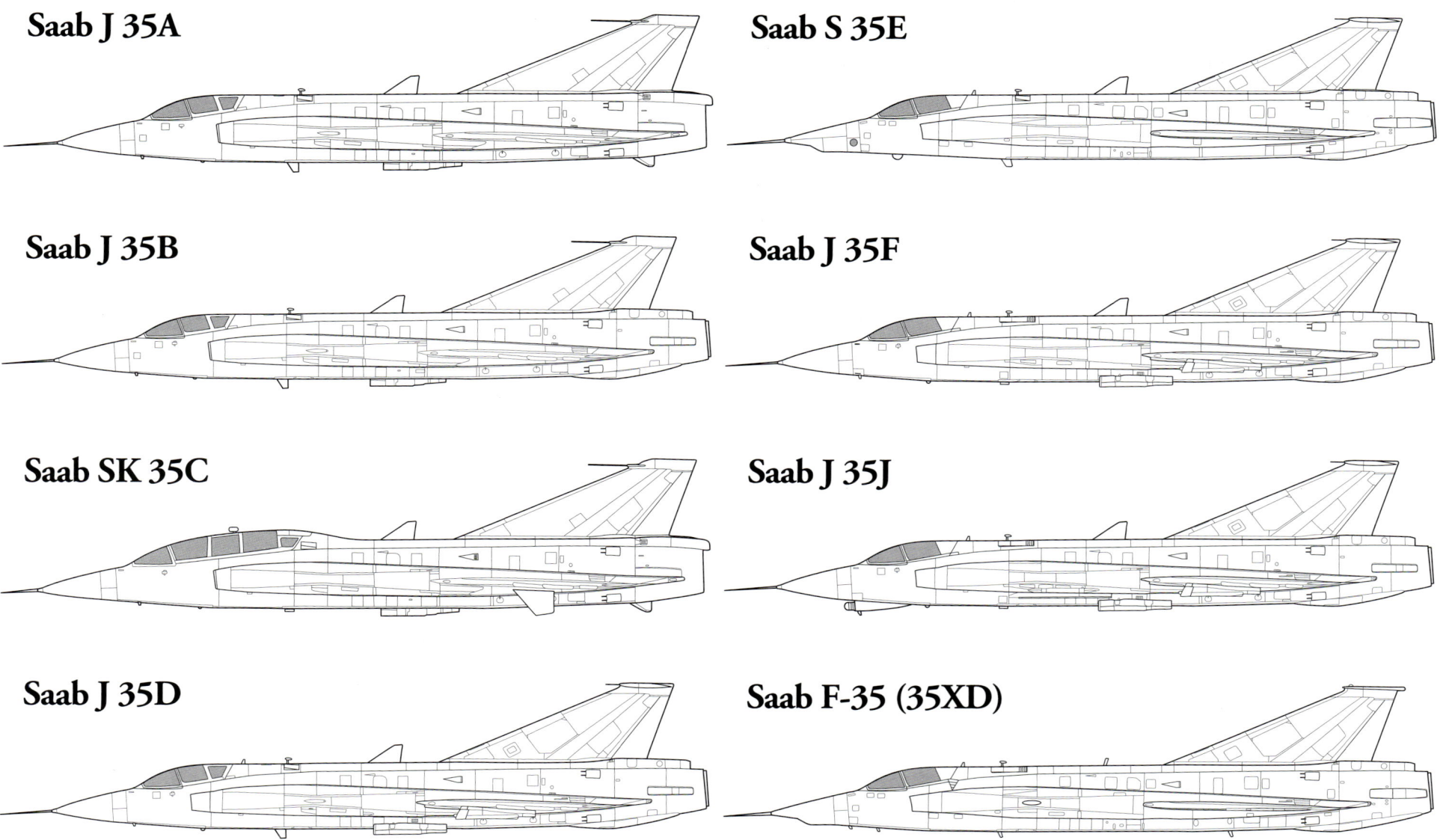# Saab 35 Production Variants

J 35A Adam - first production variant

Deliveries of the first production variant J 35A to Flygvapnet began in the spring of 1960 and were completed in 1962. In all, 90 J 35As were built (serial numbers 35001-35090) and the aircraft continued to fly with the Swedish Air Force until 1979 when the last A models were withdrawn from active service. By tradition in Sweden the two first figures in the serial number refer to the type while the last three are the number of the individual aircraft.

The first 65 production J 35As, up to serial number 35065, were powered by the RM6B (Avon Mk 48A) turbojet engine with a Swedish EBK 65 (type 65) afterburner. These aircraft received the unofficial name of "short-tail" Draken, or model 35A1.

The last 25 J 35As (35A2 serial numbers 35066-35090) were fitted with the EBK 66 (type 66) afterburner, delivering around 300kg more thrust to give a higher service ceiling and better performance. The longer EBK 66 afterburner also necessitated a longer rear fuselage on the aircraft. Wing tunnel tests conducted in the autumn of 1957 showed that the model with the long tail had a significant aerodynamic advantage. As the tailskid was not considered sufficient to protect the extended rear fuselage from damage during touchdown, two small wheels that could retract into the rear fuselage fairing were fitted in place of the skid.

In its initial stage, the J 35A aircraft - up to the serial number 35040 - lacked radar and a fire control system and were used for conversion training in the F13 Wing. Beginning with serial number 35041, aircraft received PS-02/A radar and the S6 fire control system. Basic avionics included a VHF/UHF radio, radio altimeter, transponder, PN-793/A identification friend or foe (IFF) system, and a Swedish copy of the US Lear-14 autopilot. Around the year 1963, 25 of the first production J 35As were modified to SK 35C two-seat trainer version.

Saab used this first-production J 35A, s/n 35001 as the "long tail" test aircraft. The early style emergency ram-air generator with a 10-blade turbine was installed. The outer wing of this craft features wing fences – an early anti-stall, vortex-generation modification. (Bertil Olofsson / Swedish Military Archives).

By tradition, the first deliveries of the J 35A went to the F13 Wing at Norrköping-Bråvalla in March 1960. Nineteen of original J 35As delivered to F13 Wing for service tests and evaluation are visible in this view. (SFF Photo archive).

This Saab J 35A 13-77, s/n 35077 was delivered to F13 Wing in 1962 with a complete set of electronic equipment, including an S6 fire control system and PS-02A radar. This aircraft was, accordingly, used for radar training. The demarcation line between the top and bottom camouflage colors is unusually low on this aircraft. (SFF Photo archive).

Saab J 35A F13-84, s/n 35084 takes part in winter exercises in 1962. Spread across the country, military air bases make Sweden's air force units less vulnerable to attack and allow for maximum levels of mobility. (Bo Widfeldt / SFF Photo archive)

This J 35A 16-51 of F16 Wing came equipped with the S6B fire control system, which incorporated the IR scanner. This sensor differed considerably from the more advanced Hugges S71N sensor installed on the J 35F2. (Lars Soldéus / SFF Photo archive)

A5 launching rails attach large 135mm air-to-surface rockets to hard points 1, 2, and 3. Two rockets could piggyback on a single rail. There are slots on the stabilized fins and the removed underwing pylon (hard point 4). (Swedish Military Archives)

Y-shaped pylons allow two Rb 24 missiles to be installed on the centerline fuselage pylon. Pictured here are two dummy Rb 24 missiles used for verifying mass and aerodynamic compatibility of a new pylon. (Swedish Military Archives)

This Saab J 35A F16-52, s/n 35090 was delivered to the Swedish Air Force on 11 December 1961 - the last A-type Draken delivered. After Flygvapnet service, it was moved to a museum on 12 December 1979. This photograph was taken in October 2008 when the aircraft was awaiting restoration on the territory of the Malmen Air Force base.

This Saab J 35A is fitted with the aircraft's early-style canopy and short engine inlets. The national insignia on the fuselage was 600mm in diameter at the beginning but was reduced to 450mm for J 35J variant. The Wing number '16' was placed on the fuselage and aircraft identification number '52' on the vertical fin.

The J 35A '52,' s/n 35090 belonged to the F16 Wing in Uppsala. This aircraft was one of the F16 Wing's display team. Aircraft wore standard camouflage and a small yellow stingray emblem painted on the tail fin. The F16 display team under the leadership of Captain Rolf Gustavsson performed at a number of different air shows.

The inner wing of the Saab 35 double delta had a 79.4° sweep and was thick enough to accommodate the air intakes, weapons, fuel, and landing gears. The outer portion, swept at 57°, was thinner and provided lift for short-field and low-speed operation while still keeping drag low enough for supersonic flight.

The elevon's hinge points and mass balances are clearly seen in this view of the outer elevon section. The leading edge of the wing is left unpainted.

Yaw control was maintained by using the vertical stabilizer and hydraulically-operated rudder. There is a large aerodynamic ballast atop the rudder. A pitot tube is mounted on the early-style fin, which displays the F16 Wing badge and yellow stingray marking.

The engine cone with afterburner cooling scoop are seen close up. Four quick-action locks join the "long tail" cone to the rear fuselage section, where the break line is evident.

J 35B Bertil - operational interceptor

Saab engineers began development of the new Draken version J 35B in 1956 even before the J 35A went into production. The J 35B became the first fully-operational interceptor version of the Draken. It also had a secondary fighter-bomber capability. First flown on 29 November 1959, the test B-version of the aircraft was in fact the rebuilt airframe of the J 35A, s/n 35011.

A number of changes had to be introduced but principle improvements included increased thrust from the advanced engine with afterburner, greater fuel capacity, and improved avionics. Airframe development was complete before the more powerful RM6C engines and PS-03/A radars with S7 fire-control systems were ready and therefore first-production J 35B aircraft were equipped with the old RM6B engine, PS-2 radar and C6 fire-control system. The J 35B aircraft were upgraded to full specification later.

Upgraded aircraft could be fully integrated into Sweden's semi-automatic air defense system STRIL 60. STRIL was a Sweedish acronym for "Intercept Control and Early Warning." The system, very advanced for the 1960s, provided the pilot with information regarding the heading, distance, and altitude of a target. A total of 72 J 35B aircraft (serial numbers 35202-35273) were delivered to the Swedish Air Force between February 1962 and March 1963. Retirement of the J 35B began in 1974 and was completed in 1976. After being withdrawn from the Flygvapnet, the J 35B continued serving the Finnish Air Force designated as the Saab 35BS ("S" standing for "Suomi," meaning "Finland").

Internal armament was similar to that on the J 35A - two 30mm M55 ADEN cannon, each with 90 rounds of ammunition. Two Rb 24 Sidewinder missiles were mounted on the outboard pylons and two on a centerline Y-shaped pylon. The J 35A and 35B version could mount six 135mm rockets on each wing, making a total of 12 rockets. To this capability the J 35B added two fuselage-mounted rocket pods containing 19 75mm folding-fin rockets making total load of 38 rockets.

Rockets used on 35B and 35D were designed to hit airborne targets in all weather conditions. Later, rockets mounted on the J35F/J gave the Draken ground attack capability. The original Mk 4E reflector gun sight served both air-to-ground and air-to-air sighting. Once they were upgraded to full J 35B standard, the PS-03/A radar and S7 fire-control systems greatly improved the aircraft's weapons accuracy. In later years the J 35B was fitted with the 73SE-F ejection seat system, providing improved low level, high-speed capability.

Saab J 35B 18-47, s/n 35247 of F18 Wing is seen here in 1962. The Draken adapted easily to operating from snow and ice-covered surfaces and in the numbing cold of the Scandinavian climate. (Lars Soldéus / SFF Photo archive)

This J 35B 18-47, s/n 35247 was an early-production aircraft equipped with early RM 6B engine and a limited set of electronic equipment. In June 1962, the F18 Wing began to receive the J 35B to replace the J 34 (the Flygvapnet designation of Hawker Hunters). Later aircraft were upgraded to full J 35B specification.

From left to right, J 35B weaponry includes a 500-liter external drop tank, a pocket pod (with a frangible nose cone) containing 19 75mm FFARs, an Rb 24B air-to-air missile, a 500kg bomb, a 250kg bomb, six 135mm air-to-surface rockets, three 80kg bombs, and a 30mm M/55 ADEN cannon. (Bertil Olofsson, from Swedish Military Archives).

F10 Wing's J35B 10-57, s/n 35210, wears temporary checkers on its fin and wings during tactical exercises in Ängelholm in August 1968. (Lars Soldéus / SFF Photo archive)

The F18 Wing began to replace its Hawker J 34 Hunters with new J 35B planes in June 1962. This J 35B 18-65 carries two Rb 24 missiles installed on the Y-shape underfuselage pylon. Rocket lunching rails are visible under the outer wing. (SFF Photo archive)

Some Flygvapnet bases, notably that of the F18 Wing at Tullinge, had underground hangars in which aircraft like these Saab J 35B aircraft of F18 Wing were protected even from nuclear attack. (Bo Widfeldt / SFF Photo archive)

Seen here in October 1995, the Saab 35BS DK-206, s/n 35245, of Lapland Air Command is one of the six J 35B aircraft that were remanufactured for the Suomen Ilmavoimat (Finnish Air Force) after having served with Sweden's Flygvapnet. Saab built this aircraft on 28 June 1962 and it made its first flight on 8 October 1962. Its total flight time is 1,599 hours, 41 min; flight hours in Finland - 914 hours. It last took to the air on 6 October 1995. The aircraft is now on display in the Vantaa Helsinki museum. (Jurki Laukkanen)

Saab 35BS DK-206, s/n 35245, is the second Finnish Draken to carry the DK-206 number. The aircraft number 206 was used on two Drakens. The first DK-206, s/n 35266 was damaged by fire and withdrawn. Another ex-Flygvapnet J 35B, s/n 35245 was bought and refurbished in Finland at the Valmet Works in Kuorevesi as a new 35BS DK-206. As a result, there are photographs of two different aircraft bearing the same number.

The Saab 35BS DK-206 aircraft is on display in the Vantaa Helsinki museum. Finnish 35BS lacked radar and missile armament and were used for personnel training and technical education. The primary fighter armament consists of two wing-mounted 30mm M/55 ADEN cannon, each with 90 rounds of ammunition.

Drakens have a pressurized and air-conditioned cockpit. Faired into a dorsal spine that runs the length of the airframe, the cockpit features a rear-hinged clamshell canopy. The canopy is opened and closed mechanically with a balancing spring device.

A screwed frame secures the windshield to the fuselage. The canopy-opening handle is located under the windshield. The angle of attack indicator and gun sight are visible through the Plexiglas windshield. There is a seal between the Plexiglas and metal frame.

A pivot arm at the canopy frame's aft end attached Draken's one-piece clamshell-type canopy to the fuselage. The counterbalance strut aided the pilot in raising and lowering the canopy without the aid of power devices.

This side view of the Saab 35BS DK-206 shows the "flat" framed canopy and "short" engine intakes that were key features of early Drakens. The engine air intakes were simple, with no moving parts.

The engine compressor was equipped with inlet guide vanes to direct the incoming air, as seen in this look into the left air intake. The powerful RM 6B engine offered the aircraft minor improvements in climb, high-altitude performance, and high thrust-to-weight ratio.

The composite air intake construction was very advanced for early 1960s. The intakes are separated from the fuselage surface to avoid boundary-layer air, which may reduce engine performance.

On the left side console are cockpit lighting controls, canopy jettison knob, and non-emergency canopy release lever, on the lower left of the console are fuel and throttle controls, engine ignition switches, landing gear position lights, and the radio panel.

Appearing on the right side console are (left to right) weapons control panel, systems circuit breaker panel, fuel indicator test switch, navigation lights switch, cockpit air conditioning switches and emergency generator switch. On the lower right of the console from left are: nose wheel steering wheel and autopilot control panel.

The 35BS main instrument panel is quite Spartan. The radarscope was replaced by an artificial horizon. Prior to delivery to Finland, PS-03/A radar and Swedish navigation/landing equipment PN-549/A and PN-793/A were removed. The Saab 35 cockpit was not very roomy by today's standards but was well designed and ergonomic.

Three rear-view mirrors are installed on the forward canopy frame. A canopy handgrip is mounted on the top of the frame.

The Mk 4E reflector gun sight fitted to the 38BS was usable as a gyro sight and served for both air-to-ground and air-to-air sighting of weaponry. Imagery projected onto the angled glass.

The early-style stand-by magnetic compass at the bottom right of the canopy frame is still carried on the today's jet fighters. Under the compass is a deviation table.

The front part of canopy frame is seen in raised position. There is a seal around the edge of the canopy frame to maintain cockpit pressure when the canopy is locked.

The leather-covered headrest accommodates the helmet and back of the pilot's head in high-G evolutions. In a feature pioneered by this aircraft, the pilot's seat was angled back 25 degrees, giving the pilot more g-resistance.

The top of the ejection seat and details of the underside of the Draken's canopy are visible here. On the canopy is a pocket for the aircraft log book and behind it the serial number is painted in black. A safety pin on top of the seat secures the emergency seat. Another safety pin stored in the pocket beside the headrest was used for maintenance. The seat was automatically safetied when canopy was opened and armed when closed. VARNING KRUT RÖR EJ MEKANISMEN translates as WARNING GUN POWDER DO NOT TOUCH THE MECHANISM. RAKETSTOL - EJECTION SEAT.

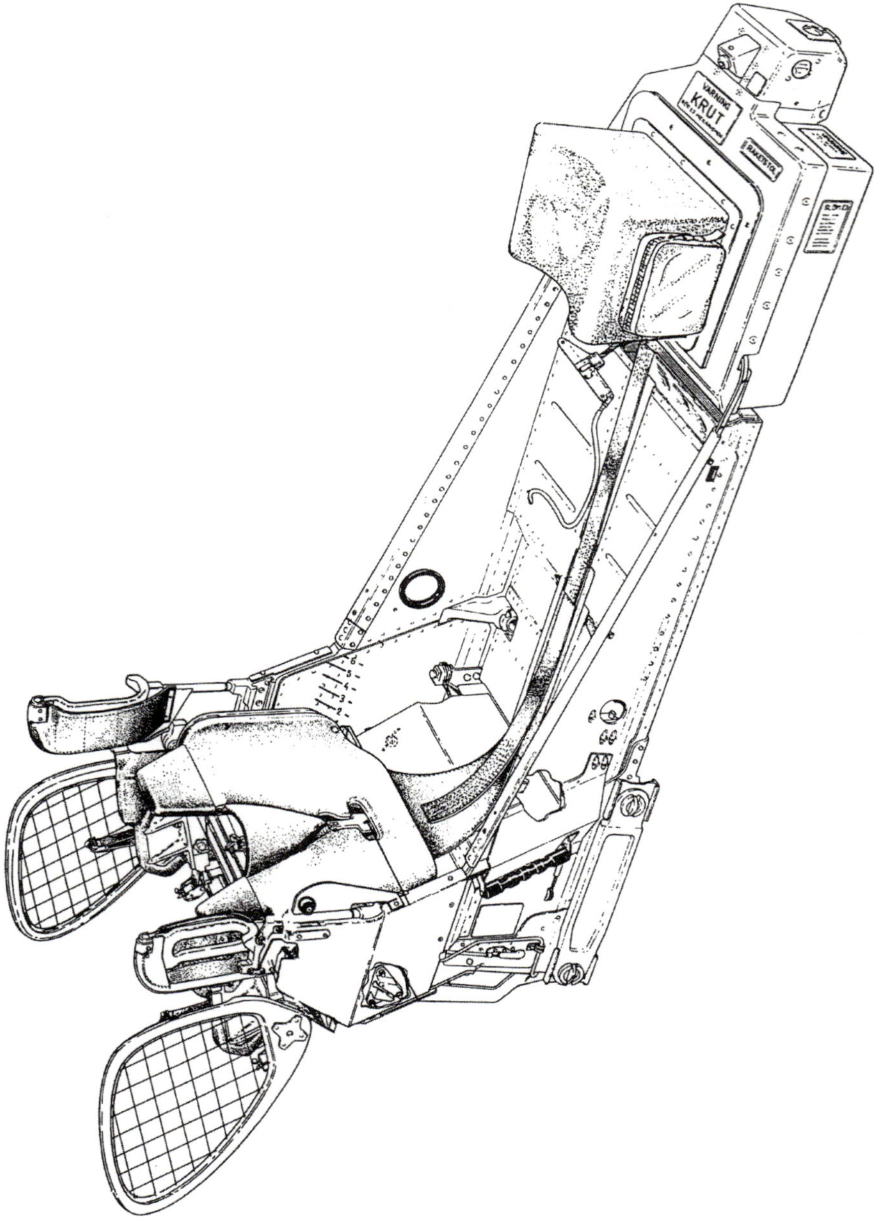

Sweden produced its own-designed ejection seat for Saab aircraft. This trend was only broken when a Martin Baker seat was specified for the JAS 39 Gripen. The Drakens were equipped with the fully automatic Saab 73SE-F rocket-assisted ejection seat, which is associated with the British GQ parachute system. The ejection seat was rated safe for ejections down to 100km/h airspeed at zero altitude. Construction was quite typical of the late-1960s. It was made of framed and riveted aluminum. The seat had special leg supports and an energy-absorbing headrest. The seat belts were integral with the parachute harness. A leg resistant system was provided to prevent leg injuries during ejection. The pilot also had to strap his boots to the seat. The legs were pulled against the seat's leg support during ejection. (Photo Jyrki Laukkanen)

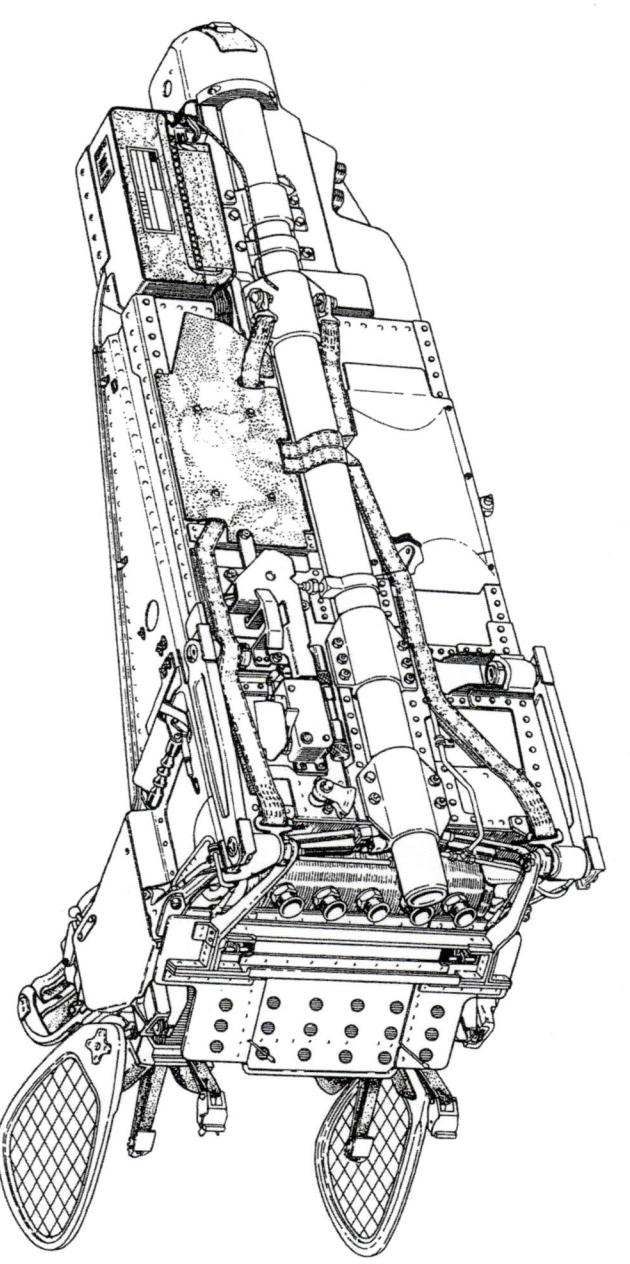

The pilot could initiate ejection by pulling either of the red ejection control handles at the forward sides of the seat. After ejection, the seat and pilot separated automatically; both came down to earth on their own parachutes. The seat-man separator was mounted behind the seat shoulder harness and consisted of a belt behind the pilot that stretched and ejected the pilot from the seat. At high speeds there was a time delay before the separation. The small parachute attached to the back of the seat deployed after the seat had cleared the aircraft and slowed the seat down until the pilot separated from the seat. The main parachute did not open until pilot had descended 3,000 meters. The bottom of the seat housed the survival pack, including an inflatable lifeboat, small food rations, and equipment the pilot needed for survival until rescued. (Photo Jyrki Laukkanen)

The landing gear is of tricycle type with additional tail bumper. The front leg, which is retractable into the fuselage, is steerable. The main undercarriage has a single braking wheel, which retracts outward. The wheelbase is 4.0m and the track of the main wheels is 2.7m. The fairly heavy landing gear strut was developed with heavy landings in mind.

This technical manual picture shows the nose gear retracting mechanism. The landing gear doors opened and closed by linkage assembly activated by the gear strut.

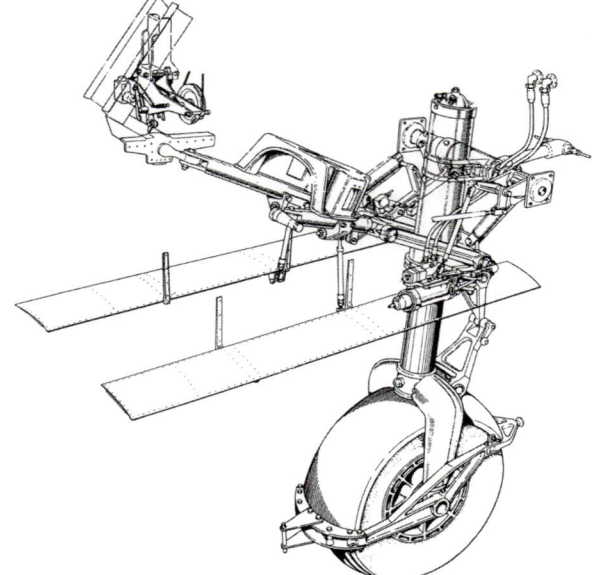

This left view of the nose gear shows the single wheel oleo strut with axle-supporting fork, torque scissor linkage, mudguard, hydraulic steering cylinder, the gear leg locking bracket, and a small handle for disconnecting torque linkage before towing.

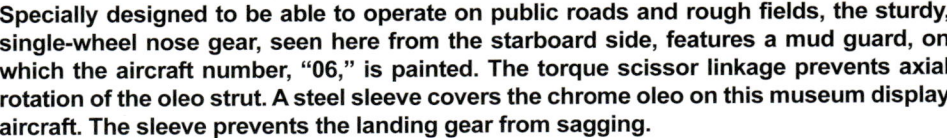

Specially designed to be able to operate on public roads and rough fields, the sturdy, single-wheel nose gear, seen here from the starboard side, features a mud guard, on which the aircraft number, "06," is painted. The torque scissor linkage prevents axial rotation of the oleo strut. A steel sleeve covers the chrome oleo on this museum display aircraft. The sleeve prevents the landing gear from sagging.

The nose landing gear, seen from the front, is steerable (by cockpit-mounted tiller wheel) and retracts hydraulically forward into the wheel. The gear leg mechanically closes the doors. If hydraulics fail, pressurized air can lower the gear. An external electric power connector is under the cover of the box labeled "LIVSFARLIG SPÄNNING" ("LIFE THREATENING VOLTAGE"). The round connector with no cover is for systems testing.

The main landing gear is stowed in the thick inner wing part when retracted. The refueling connector is located inside the starboard main landing gear bay. The aircraft uses a single-point pressure refueling system.

Among the details of the inside of the starboard main landing gear door, seen from the side, are two hydraulic actuators, locking clamps at the bottom edge, and a "gear door closed" indicator electrical breaker installed into well bay.

These two pages give an in-depth study of the main landing gear. The single-wheel main gear is retracted outward into compartments inside the inner wing. During the main wheel retraction sequence, the strut is shortened to reduce the amount of space needed to stow the landing gear.

The pictures on this page show the port main landing gear detailing. The sturdy undercarriage was constructed to withstand heavy landings and operate from public roads and unprepared landing strips. The landing gear door has two parts, one door was mounted on the outside of the strut and another door was attached to the wing.

A picture from the Draken technical manual shows the main gear detailing and gives an insight into functioning and use of the various systems.

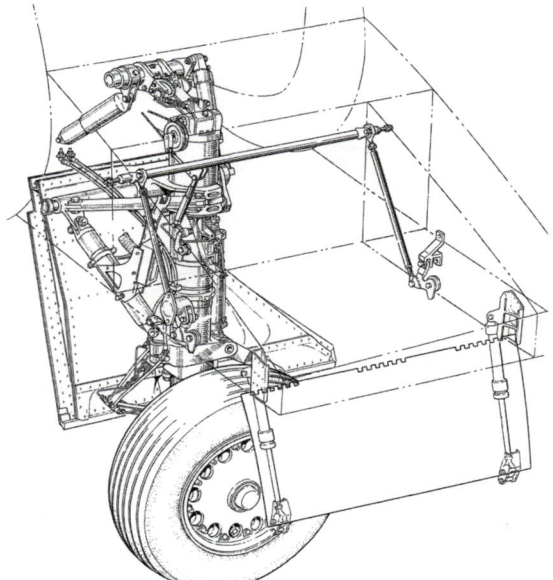

As seen in this close-up of the left main landing gear, there are 12 openings in each main wheel to allow cooling air to the brakes. The main wheel incorporates a hydraulic brake assembly and a wheel rotation sensor that are used in conjunction with the anti-skid system. Brake lines run down the side of the strut. Visible here are the torque scissor linkage, taxi/landing light, gear lock mechanism, and electrical breaker for the landing gear position indicator.

The tail wheel assembly and the air speed brake are seen from the port side. The opening in the fairing accommodates a cross bar in the hinged the tail wheel assembly. The gear doors consist of three parts. The front part is attached to the tail wheel assembly and the outboard doors are piano-hinged to the fuselage structure.

This image shows the tail wheel assembly construction features, including the oleo strut, hydraulic retracting mechanism, and wheel bay doors.

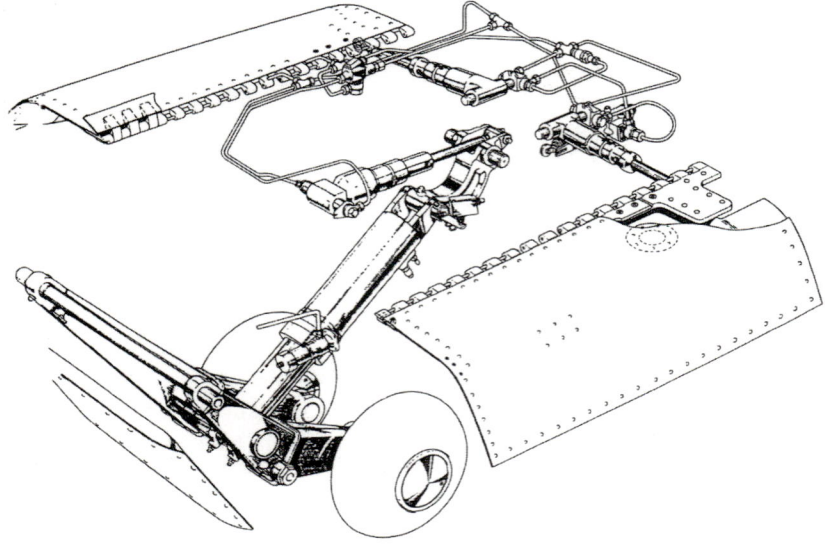

The late-production 35A and all following Saab 35 versions had a lengthened tail cone that incorporated a retractable tail wheel instead of a fixed skid.

The unique two-wheel retractable bumper tail landing gear on the Saab 35 is called the "sporre." It prevents the tail cone from scraping the ground during landing. Before touchdown the Draken's final speed is reduced to 285-290kmph. During basic training, pilots were taught to land on main wheels with very little flare, in a procedure similar to a carrier landing. Experienced pilots could land at low speeds on three wheels, including the tail wheel, even on slippery runways. Aircraft rolled out with high angles of attack to benefit from the aerodynamic breaking effect. At 160-200kmph, the nose was lowered and normal braking initiated.

Each tail wheel is supported by its own strut joined by a cross bar and attached to the oleo strut. Also visible here are the shock absorber strut and the hydraulic landing gear actuator.

The cannon compartment is located behind the gun blast port and covered by a removable panel. Mounted in the wing root, 30mm cannons were easy accessible to ground crews and operational turnaround time was relatively quick.

The stenciling on the bottom of the inner wing near the gun blast port, translates as HYDRAULIC FLUID (80% DTD 585 + 20% DTD 5540) INSPECTION AND FILLING (HYDRAULINESTEEN [80% DTD 585+20% DTD 5540] TARKASTUS JA TÄYTTÖ).

Installed in the 35BS is a new combined ADF-antenna in the dorsal spine in front of the dorsal VHF-antenna. Initially of light gray color, it was later painted in antistatic dark gray. The dorsal VHF-antenna was natural metal color with a black leading edge.

Under the port wing are the anti-buffet fence and hard points numbers 1, 2, 3, and 4 for mounting a wide variety of external items. Hard points 1, 2, and 3 usually accommodated 135mm Bofors air-to-surface unguided rockets with individual A5-type launching rails.

Wing and Fuselage

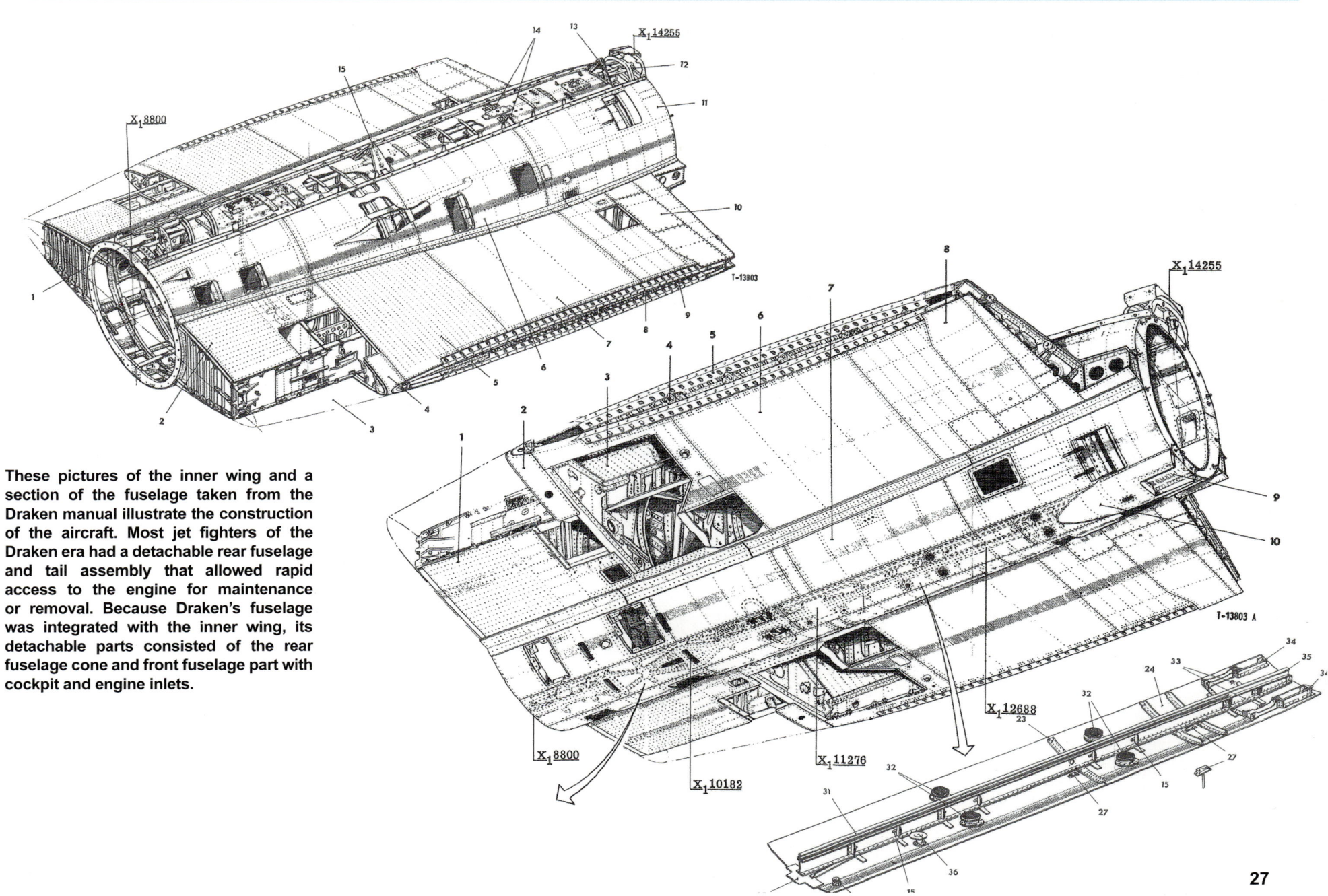

These pictures of the inner wing and a section of the fuselage taken from the Draken manual illustrate the construction of the aircraft. Most jet fighters of the Draken era had a detachable rear fuselage and tail assembly that allowed rapid access to the engine for maintenance or removal. Because Draken's fuselage was integrated with the inner wing, its detachable parts consisted of the rear fuselage cone and front fuselage part with cockpit and engine inlets.

Drakens were equipped with a ram air turbine that could be lowered into the air stream and provided aircraft systems with emergency power. This turbine was located just behind the nose gear and could be extended manually or automatically in case of hydraulic and electrical failures. In case of hydraulic failure, the ram-air power unit extended automatically and started supplying hydraulic pressure to the flying surfaces' servos (hydraulic system number two HYDR II). When electrical system failure occurred, the pilot had to turn on the emergency generator switch in the cockpit and the ram air turbine would provide aircraft systems with electrical power. In case of both hydraulic and electrical failure (engine failure) the pilot had to wait until landing was assured to lower the landing gears since electricity could fail due high hydraulic demand.

On the ground with engine stopped, the ram air turbine is always extended. It extends automatically when the engine stops and spools down. It could only be retracted when the engine was running.

A 500-liter external fuel tank for increasing combat range is attached to the centerline pylon. J 35B aircraft could only carry a single 500-liter external fuel tank on the centerline hard point. Due to Sweden's geographical location and political situation the relatively short range of the early Drakens was never a problem in Flygvapnet service.

The Finnish word "VESIKOE" stenciled on the fuselage starboard side means "WATER TEST." Fuel was checked for water content. The starter exhaust is above the pylon.

Even with an external tank mounted on the centerline pylon type T51 the aircraft was fully capable of supersonic flight. Finnish labels on the pylon read, from left to right, "GUN POWDER" i.e., ejecting the tank; "AUKI," meaning "OPEN"; and "KINNI," "CLOSED."

The aft end of the fuel tank has two stabilizing fins, added to provide the tank stability after it is dropped. The surface of the 500-liter tanks was left unpainted. The Finnish label "VEDEN POISTO" means "DRAINING WATER."

SK 35C Caesar - trainer variant

In order to simplify conversion training of new Draken pilots, a two-seater version was developed under the designation SK 35C (the SK standing for Skol, meaning Trainer). All Flygvapnet two-seat Drakens were rebuilt from the original J 35A airframes with the EBK 65 afterburner and short rear fuselage cone. The SK 35C had an extra seat for the instructor behind the student pilot, separated by a windblast screen to protect the instructor if the canopy shattered. The canopy opened to the right side. The instructor's seat was raised so he could see over the student pilot's head, but the instructor was also provided with a stereoscopic periscope to improve forward visibility. Two ADEN cannon together with the radar equipment were deleted to reduce weight.

A total of 26 of "short tail" J 35As were converted to SK 35C standard and received new construction numbers from 35800 to 35825. The conversion, carried out at the Central Verkstaden in Västerås (CVV), began in August 1961 and lasted until June 1963, by which time all SK 35Cs had been delivered. In 1966 all SK 35Cs were upgraded with the installation of PN 793/A navigation radar, new radio equipment, modification of the emergency power unit, and other changes. A small fin was added under each wing to improve yaw stability, which had been degraded by the large new canopy.

Assigned initially to the F16 Draken Conversion Unit Wing at Uppsala, the Swedish SK 35C was used during the early 1960s to provide a 25-hour conversion course to new pilots. The unique training program allowed students to begin SK 35C conversion training after only 150-200 hours of prior training on the SK 50 Safirs and SK 28 Vampire trainers. In 1986 10 of remaining SK 35Cs were transferred to F10 Wing Ängelholm. The SK 35C served until 1998 continuing to perform an important role providing operational conversion training for Swedish and Austrian pilots and "superstall" training for all Draken pilots.

This Sk 35C 16-83, s/n 35815 of F16 Wing was pictured in August 1985. A total 26 of "short tail" J 35As airframes were converted to the trainer variant in 1962 and 1963. The type performed an important role in conventional training and also served to test different weapons. It was also used as a test-bed aircraft for AJ 37 Viggen and JAS 39 Gripen Equipment.

Sk 35C '83,' s/n 35815 belongs to F16 Uppsala Wing. Like most Sk 35C, this aircraft has been left with its metal finish unpainted. Markings are minimal with F16 wing number on forward fuselage, aircraft number black "83" on the tail and wing badge. There are small fins under the wing.

This Draken, marked "800," is the first SK 35C prototype, s/n 35800 (formerly 35010). Its first flight took place on 30 December 1959. The aircraft remained with Saab and was employed as a test airframe during testing of the Saab Rb 05 air-to-surface missiles. It has tail fin markings and a modified periscope. The aircraft was scraped 1973 and ended in a fire-exercise at F3, Linköping Malmen. (I. Thuresson / SFF Photo archive)

To improve yaw stability, which had been degraded by the large, new canopy, small fins have been added under the wings of these four SK 35Cs from F10 Wing Ängelholm. They bear the emblem of Squadron 1 (1. Jaktflygdivision) "The show must go on." (Anders Nylén)

This SK 35C, s/n 35808 was rebuilt from the J 35A, s/n 35009. The drag parachute fairing is open in this picture. This aircraft was delivered to F16 Wing in 1962 and rejected due to engine blast in 1977. (I. Thuresson / SFF Photo archive)

Aviation writer and photographer Johannes Thinesen straps into the rear (instructor) seat of Saab SK 35C, s/n 35804. Pilot Hans Sundström is helping him to fasten his seatbelts in this September 1984 photo from F16 Wing, Uppsala. The windshield has a depression for the periscope. (SFF Photo archive)

Manufactured by Saab as J 35A, s/n 35020 on 21 June 1960, the Saab 35CS, s/n 35812 DK-270 of the Flight Test Center is seen in the air in March 1993. It was rebuilt as a two-seat SK 35C on 29 November 1962 and was transferred to Finland on 14 June 1984. Its last flight was on 23 August 2000. (Jyrki Laukkanen).

The two-seater has an enlarged canopy but the windshield is unchanged. The angle-of-attack indicator is visible through the Plexiglas, but the SK 35C has no gun sight.

This Saab 35CS DK-270 is on display in the Keski-Suomen Ilmailumuseo (Aviation Museum of Central Finland in Tikkakoski). The Draken training variant became invaluable in basic training and in training pilots in "superstall" recovery techniques.

The windblast screen (for protecting the instructor from a shattering canopy) can be seen through the Plexiglas of the rear part of the canopy. The periscope and instructor's rearview mirrors can also be seen.

This side view gives a good impression of the overall shape of the Saab SK 35C canopy and nose fuselage section. Instead of having two separate hoods tilting backwards, the SK 35C has one large hood hinged on starboard side.

The cooling duct for the avionics compartment is located behind the cockpit. Avionics generated considerable heat when used, requiring cooling air to ensure its normal operation.

As seen in this overall view of the windscreen and canopy, the instructor's seat was raised so that he could see over the student pilot's head. To further enhance forward visibility, the instructor was also provided with a stereoscopic periscope.

The canopy hinge points are evident in this starboard side view of the 35CS canopy. The angle of attack probe is covered. The cover protected the sensitive instruments from use or damage by unauthorized personnel.

Two-seat SK 35C (35CS) aircraft were converted from the early-production J 35A airframes and inherited the main construction features of the "short tail" Drakens equipped with the EBK 65 afterburner.

Draken's extensive tail fin and rudder, seen here from the port side, provided the longitudinal control and directional stability needed for reduced-speed landings. The rear fuselage break point is located in front of the trailing edge of the wing. This is the point where the fuselage and tail cone are separated to allow maintenance access to the afterburner.

Numerous removable panels allow access to equipment located inside the wing and fuselage. The Draken design is easy to maintain and service with minimum use of ground support units. Most of access doors and panels could be easy reached without work stands, ladders, or platforms. There are align marks on the screws.

The small fin on the dorsal spine is a VHF-antenna. The Swedes removed radio and navigation systems from aircraft slated for delivery to Finland. The Finnish 35CS was equipped with two AMR 251 VHF-radios, Collins VIR 108 VOL/ILS, Collins 860E-5 DME, Marconi AA3801 ADF, and Cossor 2720 IFF.

Air intakes, to ventilate the fuselage and reduce fire hazards, are located on both sides of the fuselage. The five pipes are part of an injector arrangement to provide cooling even when the aircraft is stationary with its engine running.

The Saab J 35A and subsequent variants have two large afterburner cooling scoops mounted on the aft fuselage tail cone. These scoops directed cooling ram air to the exhaust nozzle and aft fuselage engine bay. The scoop location on the early J 35A differed from that on late models as a result of the lengthened tail fuselage cone.

With the exception of fin top and pitot tube base, the configuration of the tail fin and rudder is basically the same on all Draken variants. The mass-balanced and hydraulically-powered rudder is hinged at its leading edge to turn to port or starboard in response to pressure on the pilot's rudder pedals. The braking parachute container is enclosed in the tailfin root. The drag parachute fairing can be seen above the "short" tail cone.

35

The drag parachute compartment was accommodated in the tail base fairing. To assist in braking, two pairs of the hydraulically actuated airbrakes are hinge-attached to the fuselage empennage section near the wing trailing edge. The aircraft serial number and warning marking are visible near the speed brake.

The blunt trailing edge of the elevon and the tail navigation light under the air scoop are visible in this view of the starboard rear of the aircraft.

To prevent engine surge at low rpm, air is passed out through this compressor air outlet located on the starboard fuselage. An air bleed valve at the eighth stage of the compressor allows air in the forward part of the compressor to pass out. The valve closes at 85% of takeoff rpm.

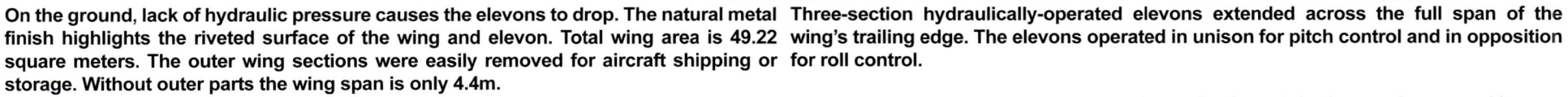

On the ground, lack of hydraulic pressure causes the elevons to drop. The natural metal finish highlights the riveted surface of the wing and elevon. Total wing area is 49.22 square meters. The outer wing sections were easily removed for aircraft shipping or storage. Without outer parts the wing span is only 4.4m.

The SK 35SC had a tailskid bumper to protect the rear fuselage during landing. On long-tail Drakens, a retractable tail wheel assembly replaced the skid.

Three-section hydraulically-operated elevons extended across the full span of the wing's trailing edge. The elevons operated in unison for pitch control and in opposition for roll control.

The bottom air brake is in the open position. The four air brakes on the top and bottom of the Draken's rear fuselage were actuated by using a switch on the throttle quadrant. Usable at all speeds, they were normally employed during approach and landing.

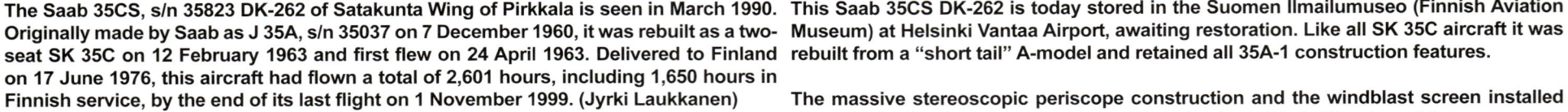

The Saab 35CS, s/n 35823 DK-262 of Satakunta Wing of Pirkkala is seen in March 1990. Originally made by Saab as J 35A, s/n 35037 on 7 December 1960, it was rebuilt as a two-seat SK 35C on 12 February 1963 and first flew on 24 April 1963. Delivered to Finland on 17 June 1976, this aircraft had flown a total of 2,601 hours, including 1,650 hours in Finnish service, by the end of its last flight on 1 November 1999. (Jyrki Laukkanen)

Two bright red triangles below the 35CS canopy are ejection seat warning stencils.

This Saab 35CS DK-262 is today stored in the Suomen Ilmailumuseo (Finnish Aviation Museum) at Helsinki Vantaa Airport, awaiting restoration. Like all SK 35C aircraft it was rebuilt from a "short tail" A-model and retained all 35A-1 construction features.

The massive stereoscopic periscope construction and the windblast screen installed above the instructor's instrument panel are clearly visible in this view of the rear canopy of a Finnish 35CS.

This photo illustrates the top of the tailfin configuration including the pitot tube base found on early Drakens.

The tail fin intake is seen here up close.

The fin houses the rudder servo mechanism, generator drive, and autopilot electronics. The intake in the leading edge of the tail fin provides cooling air to the generator.

Tails and afterburner cooling scoops are illustrated here. The first prototypes had very short and angled tail cones

The J 35A, serial numbers 35001-35065, were of the "short tail" 35A1-type design.

Late-production A and all following models featured the "long tail."

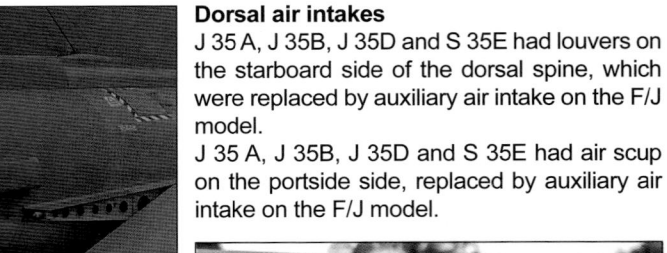

Dorsal air intakes
J 35 A, J 35B, J 35D and S 35E had louvers on the starboard side of the dorsal spine, which were replaced by auxiliary air intake on the F/J model.

J 35 A, J 35B, J 35D and S 35E had air scup on the portside side, replaced by auxiliary air intake on the F/J model.

Variations in tail fin tops are seen above. The A, B, and C models featured the early-type tail fin top, while the late-type tail fin top was standard on the D and F/J models.

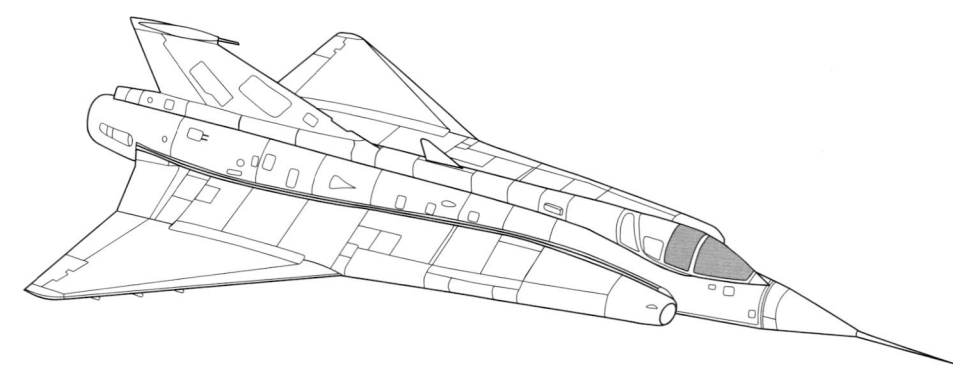

Canopies
J 35A, J 35B and J 35D flat canopy
J 35F/J, F-35 and J 35OE bulged canopy
S 35E serial 35902-35911 and 35932-35960 Flat D type canopy
S 35E serial 35912-35931 bulged F type canopy
SK 35C, TF-35 two-seat canopy

Engines and afterburners
J 35A1 35001-35065 RM6B engine with EBK 65 (type 65) afterburner
J 35A2 35066-35090 RM6B engine with EBK 66 (type 66) afterburner
J 35B RM6B engine with EBK 66 (type 66) afterburner
SK 35C RM6B engine with EBK 65 (type 65) afterburner
J 35D RM6C engine with EBK 67 (type 67) afterburner
S 35E RM6C engine with EBK 67 (type 67) afterburner
J 35 F/J RM6C engine with EBK 67 (type 67) afterburner

Pylons
J 35A, J 35B hard points 1, 2, 3 and 4 (outer wing panel) and hard point 5 (center line pylon)
J 35D and J 35F two additional underfuselage pylons hard point 6
J 35J two additional underwing pylons hard point 7
F-35 two additional underwing pylons hard point 8

Nose
S 35 E was basically the same as J 35D, unarmed with cameras installed in inner wing and redesigned nose.

40

J 35D David - more powerful Draken

The Saab engineering staff worked constantly to improve the performance of the Draken. A new variant, designated J 35D, had an improved RM6C engine, together with the new EBK 67 afterburner. It provided substantially more thrust than the original engine and delivered 57.32 kN of thrust dry and 77.28 kN kg with afterburner. The J 35D offered significant operational improvements in climb performance and acceleration. Internal fuel capacity increased by 600 liters and two additional 500-liter tanks could be carried on two new side-by-side pylons installed under the fuselage. The main external differences between the A, B, and D versions were longer, slenderer engine intakes and a redesigned fin top with a Pitot tube. These were also features on the later E, F and J variants. Electronic equipment included the S7A fire-control system, the PS-03A radar, and 05 autopilot with Mach holding mode.

The J 35D prototype flew on 27 December 1960. Serial production started in 1962 and delivery to F13 wing began in the following year. A total of 120 J 35D aircraft (serial numbers 35274-35393) were built, and they remained in operational service until 1984. Early-production D aircraft lacked a complete set of electronic equipment and had to be modified later. Those modifications incorporated not only updated electronic equipment but also a new ejection seat. In the year 1966, 28 aircraft from the first production batch were returned to Saab and converted to the S 35E reconnaissance version. Twenty-four ex-Flygvapnet J 35D aircraft were selected for modification to J 35OE in 1987 and exported to Austria.

The fourth Draken service version is the J 35D. This Saab J 35D 4-16, s/n 35321 appears at Östersund-Frösön in 1974. The F4 Wing used the J 35D until 1984, when it was replaced by Saab 37 Viggens. (Thomas Welander)

Saab J 35D '16,' s/n 35321 belongs to the F4 Wing in Östersund-Frösön, in 1974. The aircraft is wearing temporary yellow checkerboard markings applied over standard camouflage during Flygvapnet tactical exercises.

Saab J35D aircraft of F13 Wing are seen at Norrköping-Bråvalla in June 1964. The aircraft numbers suggest that these J 35Ds were from the third and final production batch of this variant. (Rune Rydh / SFF Photo archive)

A technician installs a parachute in the open drag parachute housing compartment. With the drag parachute deployed, the Draken's landing run was in the range of 500m-600m. (Åke Andersson / SFF Photo archive)

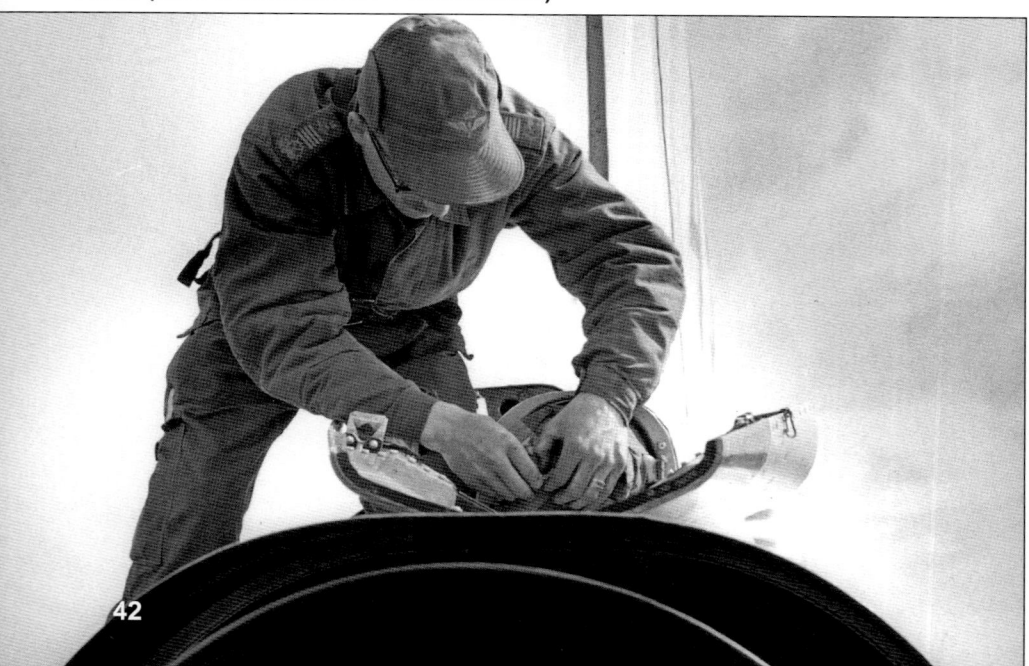

By tradition, the F13 Wing is the first to receive new fighter types. Two early-production J 35Ds, s/n 35301 and 35304, appear here in a 1963 training flight, each with four Rb 24 air-to-air missiles on wing and fuselage pylons. (I Thuresson / SFF Photo archive)

To ensure the survivability of its fighters, Sweden used various dispersal and protection methods. This F3 Wing J 35D was hidden under a camouflage net in a backwoods facility during Flygvapnet exercises in 1971. (Åke Andersson SFF / SFF Photo archive)

This Saab J 35D 13-44, s/n 35344 of F13 Wing is being prepared for its next flight. The Draken was designed to be "turned around" between missions in no more than 10 minutes. The fuel hose is connected to refueling point in the landing gear bay. (Åke Andersson / SFF Photo archive)

Mechanics load the 30mm ammunition belt. Large removable gun bay panels provided easy access to wing-mounted cannons. A key factor in the Draken's effectiveness was the fact that it could be serviced anywhere with all the necessary equipment being transported by truck. (Åke Andersson / SFF Photo archive)

Ground crew refuel the J 35D 13-65, s/n 35365 of F13 Wing. Like all Draken's support equipment, the 2,500-liter fuel tanks and pump were designed for rapid transport into the field. (SFF Photo archive)

Ground crew replace under-fuselage drop tanks with rocket pods, each holding 19 tubes for 75mm folding-fin air-to-air rockets. For better aerodynamics, the pods' noses and tails are covered by plastic cones. These weapons were installed on J 35B and D planes to allow strikes on targets in clouds when infrared missiles could not be used. Starter exhaust has stained the starboard wing bottom. (Åke Andersson / SFF Photo archive)

43

S 35E Erik - reconnaissance version

Development of a photoreconnaissance Draken began just as the J 35D went into production. The first S 35E test aircraft (with S standing for "Spaning," meaning "reconnaissance") was converted from a J 35D airframe and made its maiden flight on 27 June 1963.

Cannons, radar equipment, and related fire controls were deleted to accommodate cameras, which were manufactured by French OMERA/Segid company and developed to Swedish specifications. Five cameras were installed in the reshaped nose:
- a long focal-length forward-looking SKA 16B camera,
- a downward vertical looking long focal-length SKA 24-600,
- two sideways-looking, long focal-length SKA 24-100,
- a portside wide-angle SKA 24-44 camera.

Two sideways-looking, long focal-length SKA 24-600 cameras, which focused through periscopes, were mounted inside the wing gun bays. For easy access to the cameras, the nosecone could slide forward on rails. The pilot had a vertical Junger sight to help aim the aircraft/cameras and a cockpit voice tape recorder for saving pilot comments on the imagery.

The S 35Es were hardly modified during their service. In the early 1970s the Flygvapnet's S 35Es were upgraded and given an improved film/camera suite and the ability to carry the Vinten Blue Baron multi-sensor night-photography pod on a centerline pylon. They also had more sophisticated counter measures including chaff/flare dispensers attached to the sides of the engine exhaust and two small jammer pods on the outer wing pylons.

Of the 60 S 35Es produced, two were prototypes and 28 converted from J 35D airframes. The S35Es were operated by three Reconnaissance Squadrons: the 1 and 2 Spaningsdivisions of F11 Wing at Nyköping-Skavsta, and the 1 Spaningsdivision of F21 Wing at Luleå-Kallax. The S 35Es served the Flygvapnet until 1979 and were replaced by SF/SH 37 Viggens.

This S 35E 11-53, s/n 35953 aircraft was converted from an early-production J 35D airframe s/n 35293 and kept the original "flat" canopy. (SFF Photo archive)

S 35E 11-51, s/n 35951 of F11 Wing cruises at low altitude during a routine sortie. For many years the Draken conducted reconnaissance almost exclusively at very low altitudes because this was believed safest for aircraft operating without fighter cover. (Rune Rydh / SFF Photo archive)

S 35E 11-47, s/n 35947 takes part in combat exercises at Stigtomta in 1971. The Flygvapnet used dispersed air bases and strengthened and widened sections of public roads for flying operations. (SFF Photo archive)

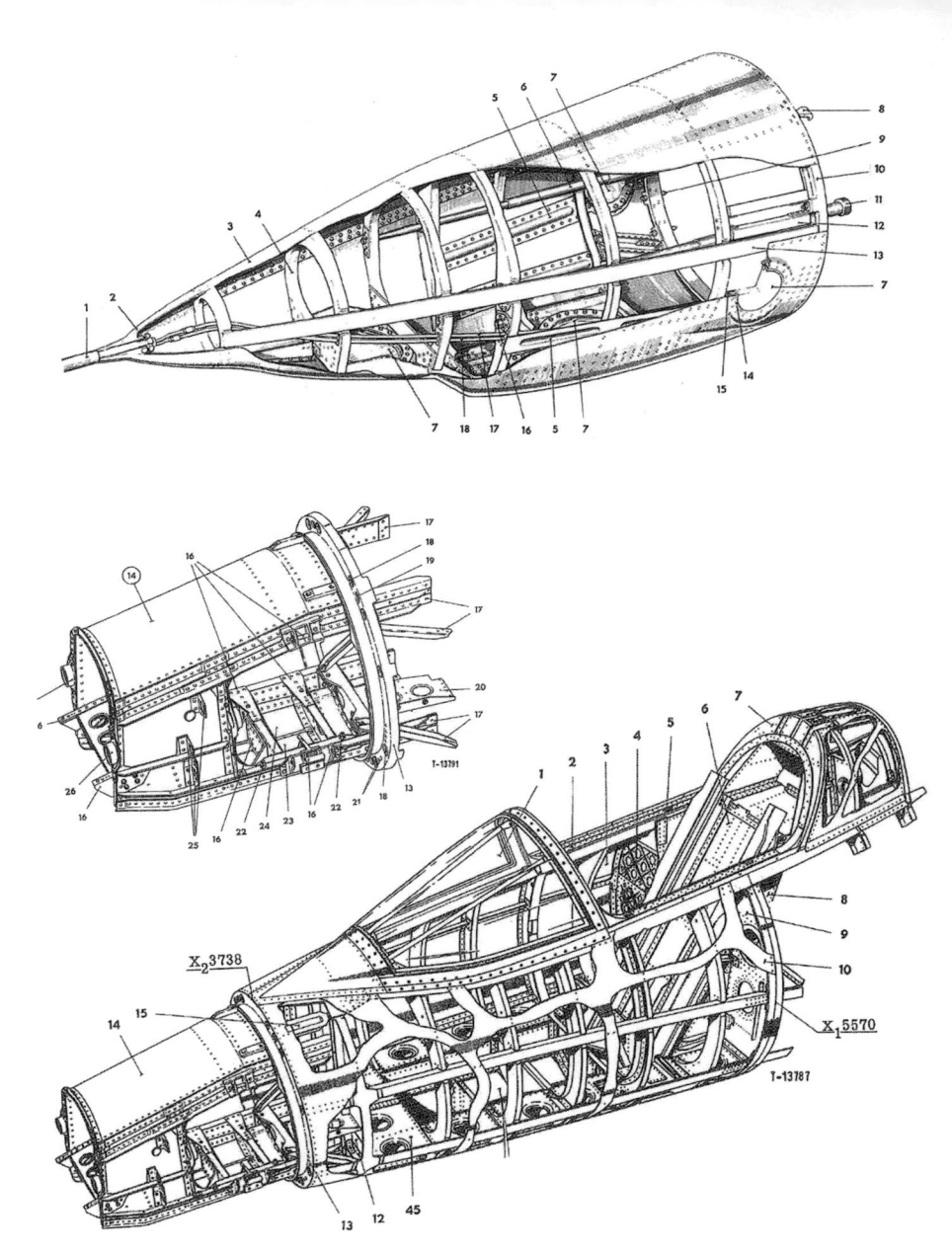

S 35E 11-51, s/n 35951 of F11 Wing flies over its Swedish homeland. Operational photoreconnaissance sorties were usually flown with four external fuel tanks. There are camera ports in the wings. (Rune Rydh /SFF Photo archive)

S 35E 11-47, s/n 35947 of F11 Wing undergoes service at an individual parking area at one of Sweden's dispersed combat bases during an exercise at Stigtomta in 1971. The nose section is moved forward to give the technician access to the camera bay and film magazines. The side-looking oblique camera mounted on the hinge frame is lowered. (SFF Photo archive)

Apart from the addition of the cameras and deletion of the radar and cannons, the S 35E was the same as the J 35D. To accommodate photo equipment, the Draken's nose and nosecone had to be rebuilt. These pictures from an S 35E technical manual show the construction features of the nose and front fuselage section.

J 35F Filip - definitive Draken fighter

The Saab J 35F was a much-improved version of the Draken, incorporating a new RM6C engine and, more importantly, new air-to-air missiles, and updated avionics. The principle improvement was the Saab S37B collision-course fire-control system coupled to Ericsson PS-01/A radar. Other upgrades included the installation of the PN-594/A navigation radar and the PN-793/A IFF (Identification Friend-or-Foe) equipment, updated navigation, autopilot and communications systems. The fire-control system was integrated with the STRIL 60 semi-automatic air-defense system. In its role as a ground-controlled interceptor, the aircraft, if already on alert, could be airborne in about 30 seconds. As regards combat air patrol, two of the new Drakens provided the airborne surveillance element of an early warning network.

The 35F carried a new weapons suite, including the Rb 27 or Rb 28 anti-aircraft missile systems. The Rb 27 and Rb 28 were Swedish versions of the American Hughes AIM-4 Falcon, featuring several improvements on the U.S. missile. The port cannon was deleted, leaving the J 35F with only one ADEN M/55 with 90 rounds of ammunition in the starboard wing. Maximum weapons load was considerably increased to 4,082kg.

The first J 35F prototype was based on the 35-10, s/n 35110 airframe (which, in turn, had been converted from J 35A, s/n 35082), and first flew on 22 December 1961. Later, other test aircraft (the Saab 35-6, 35-7, 35-8, 35-9, 35-11, 35-12 and 35-13) joined the intensive test program. On 26 June 1964 the first production J 35F took off on its maiden flight. The first wing to get new aircraft was F13 at Norrköping-Bråvalla, as was traditional, but soon a total of 18 squadrons were flying the Saab J 35F, making it the most widely used Draken variant.

Saab developed electronic equipment in parallel with the output of the first 100 J 35Fs. The result was an updated F version with the S71N IR Sensor, designed by Hughes and license-built by Ericsson. This new F variant was designated with a "2" suffix. Between October 1967 and June 1972, when Draken production finally ceased, 130 J 35F2 aircraft rolled off assembly lines. The J 35F, the last new-build Draken in the Flygvapnet, served until 1989.

The J 35F production line at Saab's Linköping plant is seen in June 1967. The aircraft '473' during final assembly is J 35F, s/n 35473. A total of 230 of J 35F aircraft were manufactured and delivered to Flygvapnet. (I Thuresson / SFF Photo archive)

The J 35F2, including the F13-26, s/n 35526 shown here, carries Rb27 and Rb28 missiles and launching rails for unguided rockets. The bottom fuselage section is unpainted. (Swedish Military Archives)

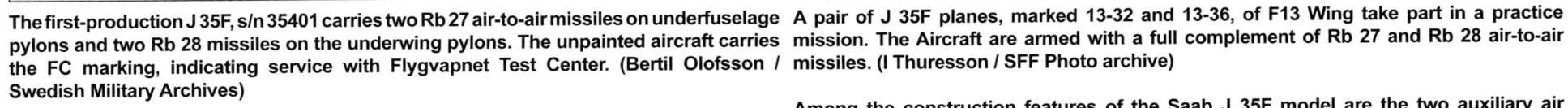

The first-production J 35F, s/n 35401 carries two Rb 27 air-to-air missiles on underfuselage pylons and two Rb 28 missiles on the underwing pylons. The unpainted aircraft carries the FC marking, indicating service with Flygvapnet Test Center. (Bertil Olofsson / Swedish Military Archives)

The Saab J 35F 12-30 armed with Rb 27 missiles taxies to a shelter. Operation from dispersed air bases continues to be the cornerstone of Swedish defense philosophy. Various kinds of protection and camouflage have been employed. (SFF Photo archive)

A pair of J 35F planes, marked 13-32 and 13-36, of F13 Wing take part in a practice mission. The Aircraft are armed with a full complement of Rb 27 and Rb 28 air-to-air missiles. (I Thuresson / SFF Photo archive)

Among the construction features of the Saab J 35F model are the two auxiliary air intakes installed behind the cockpit on the dorsal spine of this Draken belonging to the F12 Wing. Also characteristic of the 35F is the joining line between the fuselage and engine intakes. (Flygvapnet Archive)

47

Saab J-35F Draken Specifications

Engine	RM6C
Afterburner type	EBK 67
Trust (dry)	57.32 kN
Trust (with a/b)	77.28 kN
Length	15.34 meters
Wingspan	9.42 meters
Height	3.89 meters
Wing area	49.22 square meters
Empty weight	7,425 kilograms
Max t/o weight	11,914 kilograms
Max lading run	1,220 meters
Internal fuel capacity	2,865 litres
Radar	PS 11/A
30 mm M/55 canon	1
Rb 24 AAMs	2-4
Rb 27 AAMs	2-4
Rb 28 AAMs	2-4
Max speed	Mach 2.0
Climb rate	250 meters per second
Ceiling	20,000 meters
Range	2,570 kilometers

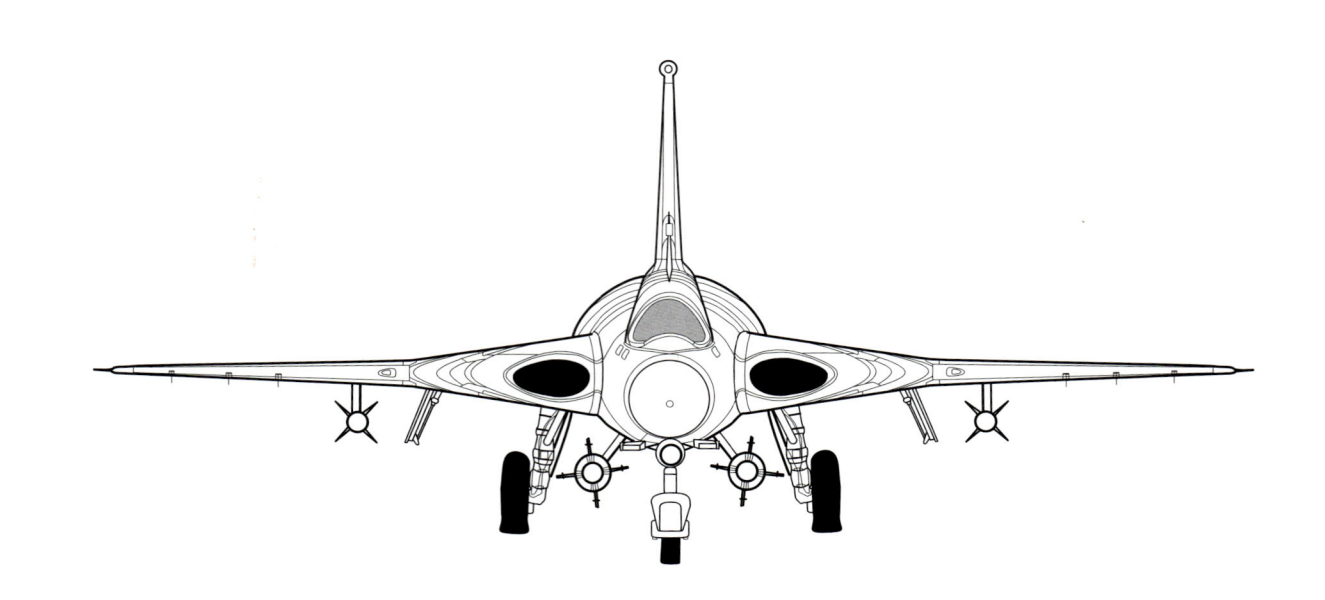

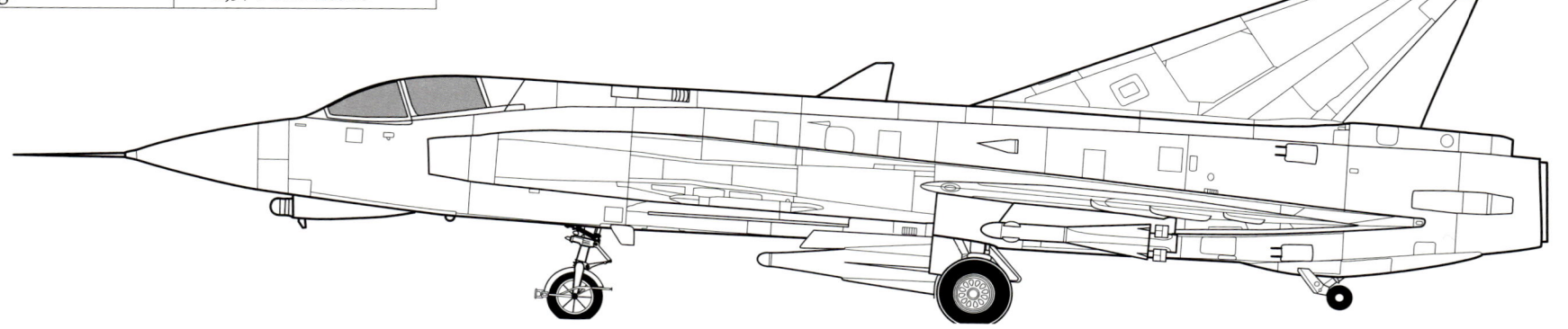

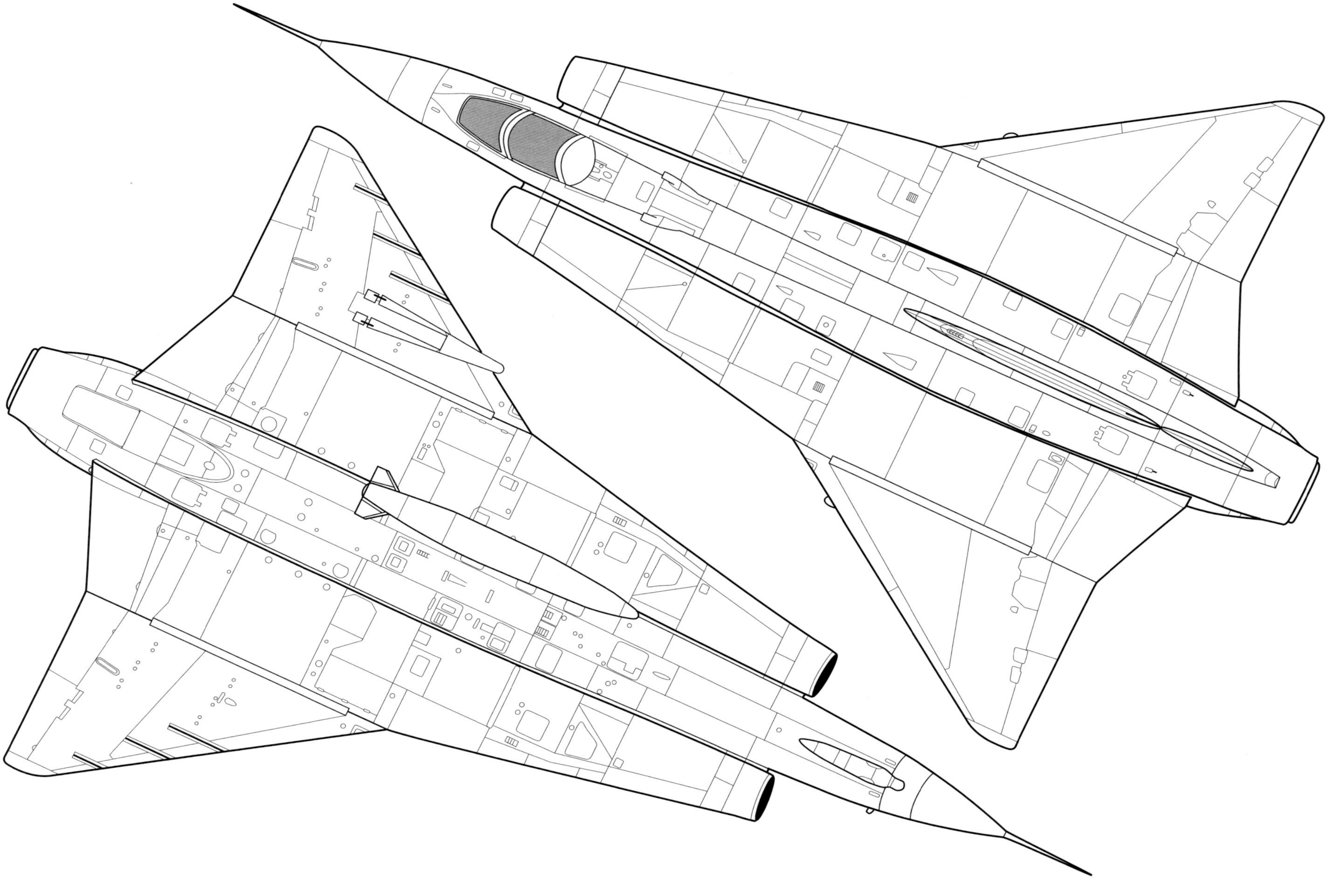

1. Altimeter
2. Accelerometer
3. Angle-of-attack indicator
4. Air speed and Mach indicator
5. Optical sight unit
6. PS-01/A radar scope
7. Distance-altitude-command indicator
8. Clock
9. Standby horizon
10. Altitude pre-select
11. Exhaust gas temperature indicator
12. Brake pressure manometer
13. External tanks in use indicator
14. Warning and caution lights
15. Switch for firefighting
16. Fuel quantity indicator 100% + 40% with drop tanks.
17. Artificial horizon
18. RPM-meter (in %)
19. Rudder pedals with toe brakes
20. Parking brakes handle
21. Control stick grip
22. Course adjuster
23. Radio magnetic indicator
24. Turn and bank indicator
25. Standby altimeter
26. Cabin pressure meter
27. Standby speed indicator
28. Speed brakes Indicator out/in
29. Landing gear warning light (speed below 500 km/h and gear not down)
30. Master warning light
31. Radio button
32. Switch for autopilot
33. Elevator trim
34. Safe and trigger for weapons
35. Quick release for autopilot
36. Vibrator for stall warning system
37. Altitude warning
(in autopilot altitude hold-mode)

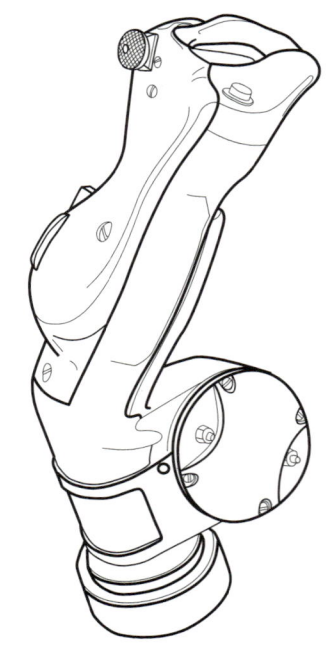

Left Side Console

1. FD 11 (system for guidance from radar controller via data transmitting)
2. Cockpit pressure regulator
3. Unit for programming FD11
4. Panel light armature
5. Cockpit light switch
6. Panel light regulator
7. Throttle quadrant
8. Instrumental panel light regulator
9. Speed brakes switch
10. Throttle friction handle
11. Canopy emergency jettison button
12. Emergency extension landing gear handle
13. MTR - engine maximum temperature regulator on/off switch
14. Radar PS-011/A steering handle
15. Radar panel
16. Afterburner switch
17. Landing lights switch
18. Canopy manual unlock handle
19. Taxi lights switch
20. Potentiometer sound strength Rb 324 missile tone
21. Potentiometer infrared searcher sound strength tone
22. Emergency lights switch
23. External fuel tanks jettison button
24. RAT switch (opens afterburner eyelets on land to cut thrust and exhaust temperature)
25. Landing gear position indicators (green lights: gear down, red lights: in transition, lights off: gear up)
26. Radio panel FR 28
27. Stall warning test switch
28. Radio panel FR 22 (FR 28)
29. Radio panel FR 21
30. Tape recorder switch
31. Afterburner fuel switch
32. Engine fuel switch
33. Ignition switch
34. Engine start switch
35. Generator switch
36. Main electrical power switch
37. Landing gear control handle
38. Drag chute handle
39. Emergency trim elevator
40. Canopy ejecting system accumulator
41. Ventilation valve for pilot isolation suit

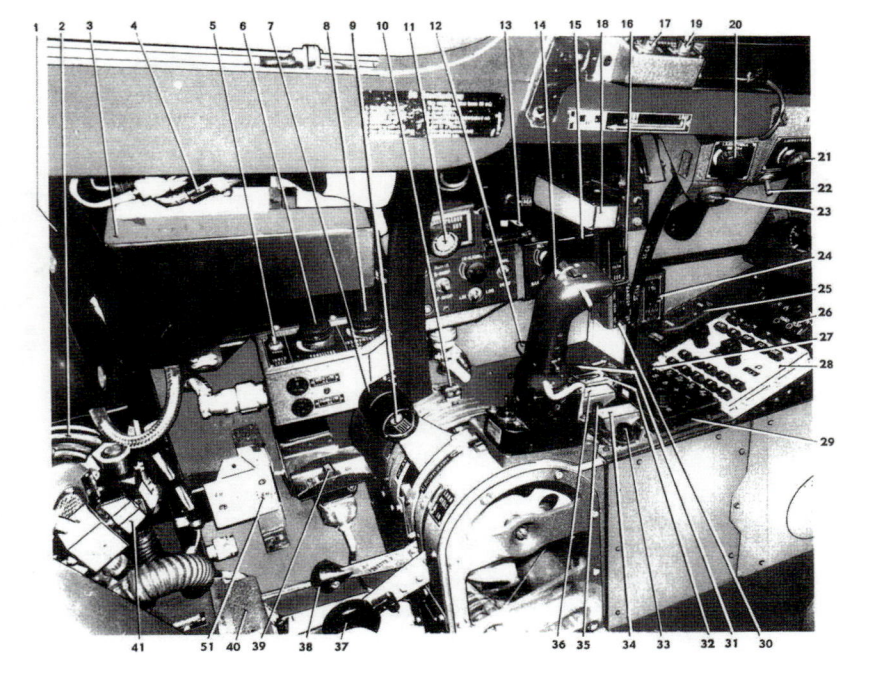

Right Side Console

1. Panel light armature
2. Drop tanks measurement switch
3. Pilot suit ventilation switch
4. Autopilot test switch
5. Weapon fire bypass for ground test
6. Weapons control panel
7. Parking holder for central coupling to oxygen/g-suit
8. Engine anti icing switch
9. Navigation lights switch
10. Circuit breakers
11. Dimming indication lamps
12. Emergency generator switch
13. Fuel system test
14. Indication lamps test
15. Connector for testing rpm-meter
16. Sparkplug switch
17. Connector for testing radar and sight
18. Radio connector
19. Radar and IR-seeker control panel
20. Air condition control panel
21. Manoeuvre panel for flight situation-instrument.
22. Canopy anti icing switch
23. Autopilot control panel. Three autopilot modes damping only, attitude and altitude hold.
24. Missile/rockets emergency firing
25. Transmitter button
26. Nose wheel steering wheel
27. Oxygen control panel
28. Autopilot test switch
29. Stall warning on/off switch
30. Test button dynamic test
31. Navigation radar PN-594/A (like ADF with DME and ILS called Anita and Barbro)

The Saab 35FS, s/n 35448 DK-241 of the Lapland Air Command is seen here in February 1996. Saab turned out this aircraft on 21 June 1966 and it first flew on 7 September 1966. By the end of its last flight on 23 August 2000, it had racked up 2,675 flying hours, of which 1,134 were in Finland, The aircraft is currently on display at the Keski-Suomen Ilmailumuseo (Aviation Museum of Central Finland) in Tikkakoski. (Jyrki Laukkanen)

The DK-241 arrived in Finland on 13 November 1986. Large fuselage letters DK (standing for Draken) and numbers were initially taped on the upper sides of the aft part of the fuselage. These national markings are also relatively large. The fuselage roundels have a diameter of 720mm with 10mm black outer outline. The wing roundels are 960mm in diameter. Finland bought 24 ex-Swedish J 35F aircraft, designated as 35FS. (Jyrki Laukkanen)

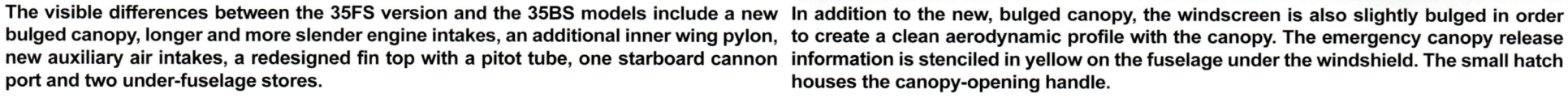

The visible differences between the 35FS version and the 35BS models include a new bulged canopy, longer and more slender engine intakes, an additional inner wing pylon, new auxiliary air intakes, a redesigned fin top with a pitot tube, one starboard cannon port and two under-fuselage stores.

This port-side view clearly shows the shape of the canopy and long engine intakes. Intake lips were normally left in unpainted fiberglass color. The yellow stenciled Finnish word "VAARA" means "danger" ("fara" in Swedish).

In addition to the new, bulged canopy, the windscreen is also slightly bulged in order to create a clean aerodynamic profile with the canopy. The emergency canopy release information is stenciled in yellow on the fuselage under the windshield. The small hatch houses the canopy-opening handle.

Three large rear-view mirrors cover nearly the entire arc of the front frame within the single-piece bulged canopy. A handgrip and stand-by magnetic compass are also mounted on the canopy frame.

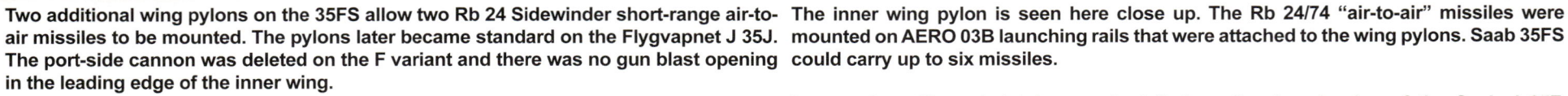

Two additional wing pylons on the 35FS allow two Rb 24 Sidewinder short-range air-to-air missiles to be mounted. The pylons later became standard on the Flygvapnet J 35J. The port-side cannon was deleted on the F variant and there was no gun blast opening in the leading edge of the inner wing.

The inner wing pylon on hard point 7 is seen here. The position of the wing pylons allowed mounting and removal of missiles without need for ladders or stairs.

The inner wing pylon is seen here close up. The Rb 24/74 "air-to-air" missiles were mounted on AERO 03B launching rails that were attached to the wing pylons. Saab 35FS could carry up to six missiles.

Improved auxiliary air intakes are installed on the dorsal spine of the Saab J 35F. Shown here is an early version of the air intake, a distinctive feature of the A, B, D and E models.

Elevons, split into three sections and extending the full length of the wing trailing edge, control pitch and roll on the Draken. The elevons operate in unison for pitch control and in opposition for roll control. The control system incorporates a three-axis automatic stability unit. The stick and rudder pedals are equipped with a synthetic feel unit.

The Draken pilot controls yaw on the aircraft by using the rudder in the vertical stabilizer. The afterburner cooling scoop location on this later Draken may be compared with the configuration on the earlier A model pictured on pages 35-36

The wing tip is attached to the end rib of the outer wing panel. Flush-head Philips screws fasten the wing tips and various panels to the airframe. These screws allow personnel to remove the panels and to access equipment located inside the wing.

Below both outer wings are three boundary layer fences, which reduce buffeting and increase pitch stability during high-angle attacks. "JORDA HÄR" stenciled on the inner fence translates as "ground here" in reference to electrical grounding of the airframe.

Inner and outer elevon sections are joined.

The 30mm M55 ADEN cannon is mounted inside the starboard wing beside the air intake. The thick inner wing section provides sufficient space for the gun and ammunition feed system. The gun port position also prevents exhaust gas ingestion by the engine air intake. Two small openings in the leading edge are cannon bay panel locking points.

The NASA-type engine cooling air intake is located near the dorsal spine. Align marks are visible around the screws holding dorsal spine and removable panel in place. Removing the panels gives access to various systems requiring regular maintenance.

Two large cooling scoops are mounted on the aft fuselage tail cone. These scoops direct cooling air to hydraulic cylinders that open and close the eyelets of the EBK 67 afterburner and aft fuselage engine bay. The aircraft serial number and warning markings are painted in yellow around the speed brake.

The ring-slot type drag parachute is housed in the fairing at the bottom of the vertical tail section. The rear part of the fairing consists of two parts that are piano hinged to the fuselage.

The starboard gun port and inner wing pylon are seen here. Added to the Finnish 35FS Draken are two external wing pylons, allowing the aircraft to carry up to six Rb 24/74 Sidewinder-class missiles on hard points 4, 6, and 7.

These early-style louvers on the starboard side of the dorsal spine are a construction feature of A, B, D, and E variants of the Draken. On the J 35F variant, seen in the photograph to the right, these louvers were replaced by an auxiliary air intake.

Auxiliary air intakes are installed on the both sides of dorsal spine. The red sign, which translates "hit here," marks a breaking point in the fuselage skin for fire fighting. There are several of these points in the fuselage through which the fire extinguishers can be punched through the fuselage skin in case of internal fire.

The starboard auxiliary air intake on the J 35F replaced the louvers on earlier Drakens. Behind the intake is a temperature probe to measure the outside temperature).

The inner wing pylon and canon port on the starboard side are seen close up.

The exhaust of the Plessey isopropyl nitrate fluid starter is located on the starboard side of the fuselage. The starter unit used high-pressure gases of ignited fluid to start the aircraft engine. The fluid tank has a capacity for three starts. The text on the sign says: "WARNING HEAVY DEVICE MOUNTED INSIDE DOOR."

Type T52 underfuselage pylons are used for carrying external fuel tanks or long-range Rb 27/28 missiles, rocket pods, and related items. This pylon is secured to fuselage hard point 6.

The angled shape of the the type T52 underfuselage pylon is apparent when seen from the front. The angle was needed to assure ground clearance.

59

Saab in Linköping shipped components to Finland, where the Valmet Works in Kurovesi would assemble the aircraft. The Saab 35S DK-233, now displayed in the Keski-Suomen Ilmailumuseo (Aviation Museum of Central Finland) in Tikkakoski, first flew on 3 October 1975. It had racked up 2,297 flight hours by the end of its last flight on 23 August 2000.

The Saab 35S DK-233, s/n 351312 of the Lapland Wing, in April 1994, was the last of 12 Valmet-assembled "Saab 35S" Drakens. The "S" stood for "Suomi," meaning "Finland."

An inscription referencing "Count Eric von Rosen" is painted on the tail of the DK-223. This Draken was officially named after the Swedish Count Eric von Rosen who donated the first aircraft to the new Finnish Air Force in 1918. Von Rosen's sister-in-law, Karin von Kantzow, married Germany's future Luftwaffe chief, Hermann Göring, in 1923.

The Valmet-assembled 35S aircraft are normally equipped with a license-built Hughes 71NS IR sensor. Aircraft were continuously upgraded so that by the time of their retirement, they had capability close to that of the Swedish J 35J. A fiberglass radome shrouds the antenna of the PS-01 fire-control radar. This ranging radar provided distance-to-target information for the fire control system.

The infrared sensor is seen close up - although the museum has reportedly replaced the actual electronic device with a mock up.

The massive C5 type pylon mounted under the outboard wing panel can carry Rb 27/28 missiles. The pylon was secured to hard point 4 in the wing undersurface.

The Rb 28 missile is seen attached to C5 underwing pylon. This infrared homing "air-to-air" missile was a Swedish-built copy of the American Hughes AIM-4 series Falcon, but featured a number of improvements compared with the American missile. Draken can carry two Rb 28/27 missiles on the wing pylons and two on the under-fuselage pylons. With the four missiles installed, the top speed of the aircraft fell to Mach 1.4.

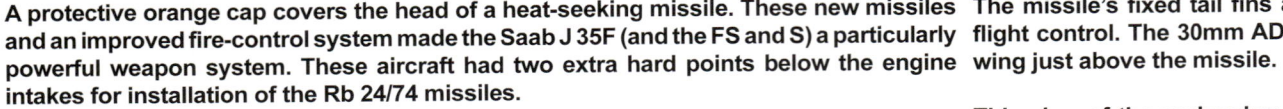

A protective orange cap covers the head of a heat-seeking missile. These new missiles and an improved fire-control system made the Saab J 35F (and the FS and S) a particularly powerful weapon system. These aircraft had two extra hard points below the engine intakes for installation of the Rb 24/74 missiles.

Standard Draken air-to-air armament consists of two Rb 24/74 missiles on G-pylons with an AERO 03B launching rail (hard point 7) and two Rb 27/28 missiles with the F5 wing pylon (hard points 4 or 6).

The missile's fixed tail fins are equipped with "rollerons" (air-driven gyro wheels) for flight control. The 30mm ADEN M/55 cannon port is visible on the leading edge of the wing just above the missile.

This view of the underwing pylon shows the missile from the fuselage side. Installed on the museum aircraft is a Soviet-built air-to-air R-13M missile (an unlicensed copy of the American AIM-9 Sidewinder). The Russian stenciling is roughly overpainted but is evident on the inner side of the missile body. Markings indicate this is an inert missile.

Chaff and flare dispensers are installed at the back of the extended afterburner cooling inlets. The Finnish Saab 35Ss were upgraded in the 1990s and equipped with chaff and flare dispensers.

Two large 1,275-liter type T52 external fuel tanks are fitted on fuselage pylons to extend the Draken's range. Two under-fuselage 500- or 1,275-liter tanks were almost a permanent Draken feature. An additional pair could also be carried on the wing pylons for ferry flights. There is also a communication blade antenna mounted under fuselage.

A look inside the exhaust of the RM 6C engine shows the afterburner flame holder and engine nozzle in the closed position. The exhaust nozzle area was automatically increased for the afterburner operation. According the Draken's flight manual, maximum use of afterburner should be no more than 10 minutes per sortie.

On the left side panel picture from L to R are: cockpit lighting controls, canopy jettison knob, and normal use lever, and below them, the radar control panel. On the lower-left console are the throttle quadrant, the radar steering handle, the interior lightning panel, landing gear position lights, part of the radio panel, altimeter and cabin pressure meter, stand-by speed indicator, and landing gear warning light.

On the right side panel from L to R are: the weapons control panel, systems circuit-breaker panel, fuel indicator test switch, external lighting control panel and cockpit air conditioning switches. On the lower-right console from left are the nose wheel steering wheel and autopilot control panel.

Dominating the main instrument panel is the centrally-mounted radar scope (with hood not fitted in this picture) and gun sight above it. Left of these, from the top, are angle-of-attack indicator, vertical tape display showing Mach number, G indicator, and altimeter. On the right are a clock, standby altitude, and EGT. At lower center from L to R are: turn and bank indicator, heading and navigation RMI, fuel gauge, and engine RPM indicator.

One of the major assets of the 35S is the Swedish Ericsson UAP 13-series fighter radar with Flygvapnet designation PS-01/A integrated with a Saab S7B collision-course fire-control system.

The PS-01 radar used in the 35S Finnish Air Force Draken is designated UAP 13104 in accordance with Finnish nomenclature.

The Saab J 35F is powered by the RM 6C (Swedish-built Rolls-Royce Avon 300 Mk. 60) and equipped with Swedish EBK 67 afterburner. The RM 6C has greatly increased thrust with maximum dry 55.41 kN and 76.05 kN with afterburner. Maximum rpm was 8,100kp.

The engine is designed for reliability and easy maintenance in harsh conditions. From 1961 through 1971 Volvo built 445 RM 6Cs under license.

J 35J Johan - final update

In the late 1980s, 67 J 35F aircraft were modified to what would become the ultimate Draken standard fighter – designated the J 35J. Although it would have been logical for the 35F version to be followed by 35G, it was decided to designate the update as J 35J because all updated Drakens would go to the 10th Wing, the last Swedish wing to operate the type, and "J" is the tenth letter of the alphabet.

The goal of the J 35J modification was to bridge the gap in air defense capability between the outgoing Draken and the upcoming JAS 39 Gripen, which would be introduced in 1996. The Draken aircraft received a comprehensive avionics upgrade, including an improved radar, fire-control system, infrared search and track sensor, navigation system, IFF, and modernized cockpit electronics. Two new stores pylons were added to each inner wing section, making a total of six. The 35J could carry four 500-liter fuel tanks as well as four air-to-air rocket pods each with 19 75mm rockets. The wing was strengthened, allowing the aircraft to easily carry its maximum ordnance load of 4,082kg. The J 35J had a slightly more powerful RM6C turbojet, developing a maximum of 79,89 kN of thrust in the afterburner. The extra thrust was certainly needed to offset the increased weights. An altitude warning system was also added. The updated J 35Js were delivered between 1987 and 1991, but in 1993 the 1st Division began operating AJS 37s, and many of the J 35Js were placed in storage at an underground facility built for the 9th Wing (F9) in Säve. The J 35J Draken had modernization potential and variety of upgrades for the Draken were studied but the aircraft became a victim not of obsolescence but of rapidly shrinking defense budgets.

Another consideration in modifying the J 35J for F10 was the desire to keep a Draken unit and its training group operational to support foreign users (Austria, Denmark, and Finland) until contracts ran out in 1995. F10 had trained Draken pilots since 1986, when TIS 35 (the type conversion unit for J 35) was relocated from F16 Uppsala. The last of student went through the course in 1996 and Flygvapnet officially retired the Draken on 12 December 1998, the final Draken sortie by F10 Wing taking place on 18 January 1999. A single Draken (J 35J 35556) was retained in airworthy condition as a part of Försvarsmaktens veteranflygverksamhet (Swedish Armed Forces Vintage Aircraft).

This Saab J 35J 10-56, s/n 35556 pictured on 26 May 2007 has been retained in airworthy condition by Försvarsmaktens veteranflygverksamhet (Swedish Armed Forces Vintage Aircraft). F10 Wing Ängelholm was the last Draken unit to operate the type and will forever be associated with the J 35J Draken that is maintained in flyable condition at the Armed Forces School (FMTS) in Halmstad. (Sven Stauffer)

Saab J 35J 10-56, s/n 35556 is finished in the two-tone air superiority gray camouflage with the yellow Flotilla numbers day-glow red call signs. The national insignia on the nose is 450mm. Fuselage bottom is in natural aluminum.

The Saab J 35J 10-62, s/n 35586 was built as J 35F2 and entered Flygvapnet service on 30 October 1969. This aircraft was one of 67 J 35F aircraft remanufactured to J 35J standard during the late 1980s. The swordfish marking on the top and bottom of the wing was derived from the emblem of the 3rd Squadron (3. Jaktflygdivision/Flottilj 10) and was applied to "No. 62" in its role as the display aircraft of the F10 Wing. The squadron's yellow color was also applied on the nose cone and top of the fin. (Anders Nylén)

This pair of J 35Js of 2. Jaktflygdivision/Flottilj 10 aircraft are finished in the two-tone air-superiority gray color applied during the type's last years in service. The picture illustrates the typical configuration of the "Johan" with two Rb 74 Sidewinders, two Rb 27 air-to-air missiles, and a pair of 500-liter external fuel tanks. The large Dayglo codes applied on the upper wing surface make the aircraft easier to see in training. These markings were to be removed in wartime. (Anders Nylén)

F10 is the last Draken Wing and the only unit to operate the final Draken fighter variant. These two gray-camouflaged J 35J Drakens have Rb 74 Sidewinders mounted on the new weapons pylons installed under the intakes of the modified aircraft. The drop fuel tanks have natural metal finish. (Anders Nylén)

Saab J 35J 10-56 Draken takes part in the Air Show in Satenas, Sweden, in 2006. The hose connected to the portside of the aircraft is piping in additional conditioned air while the Draken is on the ground. (Frank Grealish / IrishAirPics).

Draken's pilot Lieutenant P. Hedberg is seen after the flight at the Air Show in May 2007. The pilot's flight suit is the standard from the middle of 1970s. The lowest leg-pockets contain a small survival kit and emergency-signal kit. The pockets above the knees were attached with Velcro and held maps, orders, and different frequency forms. The boots had eyelets for connecting the leg-fixing ropes in the aircraft in case of ejection. The tube on the left side is from the g-suit and goes in to a central coupling where the oxygen tube to the mask also was connected. (Miroslav Patočka)

69

Danish Air Force F 35, RF 35, TF 35

Under a contract signed in March 1968, the Royal Danish Air Force (Flyvevåbnet) chose a modified version of the Draken to fulfill its strike aircraft requirements. Designated F-35, the Danish attack-fighter was similar to the Swedish J 35F but had a substantially increased load capability. Armament was mounted on nine hard points: three under the each wing and three under the belly. Unlike the J 35F, the F-35 carried two 30mm M/55 ADEN guns. The maximum ordnance load rose to 4,500kg and various external weapons could be carried, including Rb 24 and AGM-12B Bullpup missiles, rocket pods, and bombs. Two 1,275-liter external fuel tanks raised aircraft range to 3,000km. The landing gear was strengthened to handle the increased weight and a runway arrestor hook was added.

The Danish photo-reconnaissance RF-35 was similar Sweden's S 35E, with five Vinten 350 and 540 cameras in the nose. But unlike the S 35Es, the RF-35 retained two ADEN cannons in the wings, at the expense of two wing cameras. The aircraft had stores pylons, giving the reconnaissance aircraft a secondary attack capability.

The TF-35 trainer, based on the J 35F airframe with an RM 6C engine and EBK 67 afterburner, became the most powerful two-seater Draken. Combat capable, with its single 30mm M/55 ADEN cannon, it could also carry A-38K jammer pods. In the mid-1980s all TF-35s were upgraded with new weapon delivery and navigation systems (WDNS), including a laser rangefinder and marked target seeker (LRMTS) in a modified nose.

The Flyvevåbnet bought 20 F-35 ground attack fighters, 20 RF-35 reconnaissance aircraft, and 11 two-seat TF-35 trainers, as well as several S 35Es and J 35Fs for parts. Squadrons 725 and 729 at Karup Air Base flew Drakens from 1970 until 1993, whereupon the last RF35s and TF35s were withdrawn and Squadron 729 was disbanded.

In the U.S., the National Test Pilot School and Flight Research, Inc. own six ex-Danish Drakens, of which two TF-35s and one RF-35 at the Mojave Spaceport are operational.

This RF-35 AR-109, s/n 351109 was the WDNS test aircraft and made its first test flight on 17 August 1981. A Ferranti LRMTS was mounted in the nose of the craft, which the Flyvevåbnet finished modernizing at the Karup Air Base in May 1986. (Joop de Groot)

The Danish Air Force Saab F-35 A-020, s/n 351020 was pictured in November 1984. After modernization in the 1970s, the aircraft received formation lights, and fin-top and wing-tip RWR antennas. The aircraft in this picture has the standard nose cone. The nose cone was later reshaped during Weapon Delivery and Navigation System modernization (WDNS) modernization. The aircraft has six underwing stores pylons, a sensor pod, and small communication antenna rather than a vertical fin. (Rainer Mueller)

The reworked nose cone and portside canon port are evident on this Danish Air Force Saab F-35 A-019, s/n 351019 pictured on 24 August 1993. As compared with the Swedish J 35F, the Danish F-35 could carry a maximum ordnance load of 4,500kg. In early 1994 the last Danish Drakens were withdrawn from service and their duties were taken over by Flyvevåbnet's F-16s. (Joop de Groot)

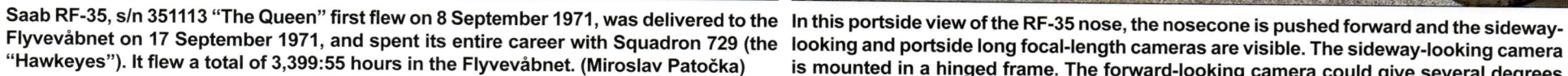

Saab RF-35, s/n 351113 "The Queen" first flew on 8 September 1971, was delivered to the Flyvevåbnet on 17 September 1971, and spent its entire career with Squadron 729 (the "Hawkeyes"). It flew a total of 3,399:55 hours in the Flyvevåbnet. (Miroslav Patočka)

The RF-35, s/n 351113 last flew on 17 December 1993 and is preserved by the local Draken Team at Karup Air Base, Jutland. A group of former and current Flyvevåbnet technicians and enthusiasts formed the Draken team in 1995 with the goal of getting the fighter airborne. (Luc Colin)

In this portside view of the RF-35 nose, the nosecone is pushed forward and the sideway-looking and portside long focal-length cameras are visible. The sideway-looking camera is mounted in a hinged frame. The forward-looking camera could give several degrees of obliquity and angle markings are evident. (Luc Colin)

Five cameras are mounted in the nose of the reconnaissance version Draken. Unlike the Swedish S 35E, the RF-35 was fully armed and did not carry cameras in the inner part of the wing. Black pipes fed air to prevent camera lenses from fogging. (Luc Colin)

71

Danish Drakens feature the radar warning receiver AN/ALR-45 (AN/ALR-69), its installation comprising six new antennas on the fin, on the nose, and in the each wing tip. (Luc Colin)

Mounted on the RF-35 outboard wing panel are formation lights, radar homing, and warning antenna. Formation lights are also mounted on tail fin, inner wing part, and dorsal spin. The pilot could adjust the pale yellow electro-luminescent lights to provide variable brightness levels during night and low-visibility conditions. (Luc Colin).

The F-35 is the heaviest Draken version built. Its landing gear was strengthened in order to bear the added weight, and a runway arrestor hook was added to the aircraft. Saab intended to add an arresting hook to all late-production Drakens but this improvement was never introduced on Swedish Flygvapnet aircraft. (Luc Colin)

The clean appearance of the emergency arrester hook on this RF 35 suggests that it hardly ever served its intended purpose. An emergency arrester hook was installed under the rear fuselage of all Danish F 35, RF 35, and TF 35 Drakens. (Luc Colin)

A National Test Pilot School TF-35 N168TP lands at Mojave, California, in the United States, after a super-stall training mission. This ex-Royal Danish Air Force AT-154, s/n 351154 was equipped with anti-spin parachute mounted in the top of the tailcone. The anti-spin chute is a safety device in case the super-stall results in an otherwise unrecoverable spin. (Alan Radecki)

The Raytheon towed decoy system prepares for a test flight of N168TP in December 2000. An electronic pod has been installed on the additional wing pylon. Danish Drakens had six wing hard points and a reinforced wing, allowing increasing load capacity.

A Raytheon towed decoy is mounted on the inboard weapons station of N168TP for a developmental test flight. The Draken's former gun bay is visible, since a panel of the wing leading edge is removed. For flight test activities, camera units and other sensors are mounted in this bay. (Alan Radecki)

Flight Research Inc. used Drakens for weapon systems tests. A Leigh Aerosystems "Longshot" adaptor mounted to an inert Mk82 practice bomb is seen ready for test flight. The "Longshot" system package included pop-out wings and a GPS guidance system that turned a "dumb" bomb into a "smart" guided, stand-off weapon. (Alan Radecki)

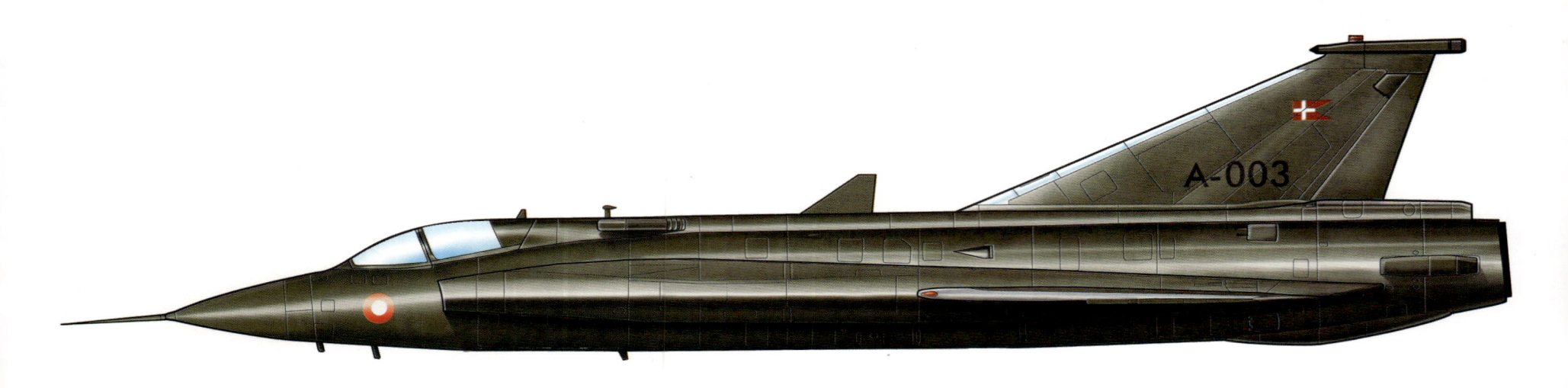

F-35 A-003, s/n 351003 was lost in an accident on 20 September 1974. On 1 September 1971 the first three F-35s (A-002, 003, 004) were delivered to Esk 725 at Karup. These aircraft were similar to the Swedish J 35F but substantially increased load capacity. The standard camouflage pattern was Olive drab, FS14079, all over. Black aircraft number and national flag on the vertical fin and national insignia were standard for this type in Danish Air Force service.

RF-35 AR-113, s/n 351113 was delivered to the Danish Flyvevåbnet on 17 September 1971 and served in Esk 729 "Hawkeyes" at Karup. It went through modification and was equipped with fin-top and wing-tip RWR antennas, chaff/flare dispensers, and formation lights. The AR-113 last flew on 17 December 1993.

Finnish 35S, 35FS, 35BS, 35CS Drakens

The Suomen Ilmavoimat (Finnish Air Force) ordered Drakens in 1970 under a deal covering 12 Saab 35S (S means Suomi) aircraft, to be assembled at the Valmet Works at Kuorevesi, as well as weapons and armament systems, spare engines, ground equipment, and other components, a flight simulator and technical documentation.

Under the initial April 1970 contract, Finland leased six refurbished ex-Flygvapnet J 35B aircraft beginning in spring 1972 for personnel training prior to delivery of the Valmet-assembled 35S fighters. The Swedes removed the navigation and landing system PN-793/A and radar PS-03/A from the planes prior to delivery to the Finns, and installed an ADF as the sole radio navigation aid. An in-flight fire in the generator of aircraft DK-206 caused its withdrawal from service on 18 January 1974. Another used J 35B, s/n 35245 was purchased and refurbished by Valmet and became the new DK-206.

In 1976 the Ilmavoimat bought six J 35F aircraft designated 35FS and three designated SK 35C (35CS). Then in 1984 Helsinki bought a second batch of 18 ex-Swedish J35Fs and two ex-Swedish SK 35Cs. The two-seat trainers allowed pilot conversion to take place in Finland instead of Sweden. In total, the Finnish Air Force bought 48 Drakens: seven Saab 35BS, 12 newly-built Saab 35Ss, 24 Saab 35FSs, and five two-seater version Saab 35CS trainers.

Procuring the Draken was important for the Finnish Air Force. When the first Saab 35S became operational, the Ilmavoimat for the first time had a fighter with an all-weather capability and radar-guided missiles. The Drakens' most active years in Lapland were from 1980 to 1995. There were enough aircraft and trained pilots for various operations, training, and tactical exercises, and combat tactics and aircraft equipment developed. Age began to take its toll in the 1990s however, and the F-18 Hornet was selected as the Draken's successor.

The official last flight ceremony took place on 16 August 2000. Six Drakens performed the last formation flight, wrapping up 29 years of Draken service with the Ilmavoimat.

Saab 35FS DK-241, s/n 35448 defends Finnish airspace in the Lapland Air Command in February 1996. (Jurki Laukkanen)

Saab 35Ss DK-223, s/n 351312 and DK-215 fly over the frozen Gulf of Bothnia in April 2000. Dk-233 pictures with a new Rb 4S IR missile. (Jurki Laukkanen).

Saab 35S DK-201, s/n 351301 of the Fight Center appears in this May 2000 photo. The aircraft was officially named "Kake" after General Kauko Pöyhönen, who was one of the key persons in the Draken procurement and was killed in a helicopter incident. (Jurki Laukkanen).

The Draken autopilot's stability augmentation system makes it easy to maneuver the Saab 35S DK-207, s/n 351304, seen in August 2000. (Jurki Laukkanen)

Saab 35CS DK-270, s/n 35812 of the Flight Test Center, seen in July 2000, is one of five two-seat trainers in Finnish Air Force service. (Jurki Laukkanen)

A special paint scheme adorns the Saab 35S DK-215, s/n 351308 in August 2000. In May 2000, before Draken's retirement, this DK-215 received a special Yellow/Black paint scheme incorporating a wisent (European bison), a feature of the badge of the 11th Fighter Squadron. (Jurki Laukkanen).